Scientific Weed Management

NIPA® GENX ELECTRONIC RESOURCES & SOLUTIONS P. LTD.
New Delhi-110 034

About the Authors

Dr. (Mrs.) T. Girija was born in Palakkad district of Kerala. She did her B.Sc. (Agriculture) at Kerala Agricultural University (KAU), Vellanikkara, Thrissur and her M.Sc. and Ph.D. in Crop Physiology at Tamil Nadu Agricultural University (TNAU), Coimbatore. She started her career in the National Seeds Project at Regional Agricultural Research station Pattambi and then worked for 20 years in All India Co-ordinated Research Project on Weed Control of Kerala as weed physiologist. She was also Principal Investigator and Head of the project for 2 years. She is a recipient of the young scientist award of the State Council for Science, Technology and Environment and also an award for best paper presentation in Biotechnology. She has specialised in the control of parasitic weed, Loranthus and published about 50 papers in peer reviewed journals. She has guided 2 Ph. D students and 9 M. Sc. students in Plant Physiology and retired as the Professor and Head of the Department of Plant Physiology, College of Agriculture, Kerala Agricultural University, Thrissur.

Dr. (Mrs.) K.M. Durga Devi was born in Thrissur district, Kerala. She acquired her B.Sc. (Ag.), M.Sc. (Soil Science & Agrl. Chemistry) and Ph.D. (SS&AC) from Kerala Agricultural University, Thrissur. She began her career as Agricultural Officer, State Department of Agriculture, Kerala and then joined Kerala Agricultural University at Operational Research Project, Ozhalapathy and later joined the All India Project on Weed Control of Kerala as residue chemist and carried out research on herbicide residues in soil, plant and aquatic systems for 25years. She handled two ad-hoc projects on "Persistence of herbicides in water bodies and its impact on aquatic system" and on "Persistence of glyphosate and POEA (surfactant) in soil and water-bioaccumulation of residues in fish" and also established a herbicide residue laboratory in the college. She has guided 2 Ph.D. students and 8 M.Sc. students in Soil Science, authored more than 50 publications and retired as the Professor and Head of the Department of Soil Science and Agricultural Chemistry, College of Agriculture, Kerala Agricultural University, Thrissur.

Dr. C. T. Abraham was born in Kottayam District of Kerala. Did his B.Sc. (Agriculture) and M.Sc in Agronomy at Kerala Agricultural University (KAU) and took his Ph.D. from Indian Agricultural Research Institute (IARI), New, Delhi. He conducted post-doctoral research at Jawaharlal Nehru University (JNU) and also at Auburn University, USA. He joined Kerala Agricultural University and was the Principal Investigator and head of the All India Project on Weed Control of Kerala for 25 years. He specialised in weed science and concentrated on weed management in plantation crops, rice and aquatic weeds. He received Fellow and Gold Medal awards from Indian Society of Weed Science (ISWS) and Crop and Weed Science Society (CWSS). He has guided 5 Ph.D. and 6 M.Sc. students on topics in weed management. He has authored and published more than 60 papers in refereed journals. He retired as the Dean of College of Agriculture, Kerala Agricultural University Thrissur.

Scientific Weed Management

T. Girija
Retired Professor and Head
Department of Plant Physiology
College of Agriculture, Kerala Agricultural University
Thrissur, Kerala

K.M. Durgadevi
Retired Professor and Head
Department of Soil Science and Agricultural Chemistry
College of Agriculture, Kerala Agricultural University
Thrissur, Kerala

C.T. Abraham
Retired Dean
College of Agriculture, Kerala Agricultural University
Thrissur, Kerala

NIPA® GENX ELECTRONIC RESOURCES & SOLUTIONS P. LTD.
New Delhi-110 034

NIPA® GENX ELECTRONIC RESOURCES & SOLUTIONS P. LTD.

101,103, Vikas Surya Plaza, CU Block
L.S.C.Market, Pitam Pura, New Delhi-110 034
Ph : +91 11 27341616, 27341717, 27341718
E-mail: newindiapublishingagency@gmail.com
www: www.nipabooks.com

For customer assistance, please contact
Phone: + 91-11-27 34 17 17
Fax: + 91-11-27 34 16 16

ISBN: 978-93-95763-87-5

Composed and Designed by NIPA®.

Preface

Weeds are integral part of agricultural production systems. They are efficient invaders and opportunistic exploiters of disturbed soils. Interference of weeds has an adverse impact on agricultural productivity, environment and human activities. They proliferate in diverse ecosystems such as wet lands, garden lands and aquatic systems. Changing climate contributes to shift in floristic composition and weed shift has become a common occurrence, adaptive mechanisms of weeds give them competitive advantage over crop plants. Weeds are endowed with superior colonizing ability, competitiveness and survival strategies, attributes that have to be understood for developing successful management technologies.

Managing weeds implies the use of various control measures which help to decrease the harm caused by weeds to an economically acceptable level. Introduction of new technologies have contributed to the emergence of new challenges in this area. While herbicides have revolutionized weed control over the years, emergence of herbicide resistant weeds has thrown up new challenges to the farmer.

This book has been envisaged to give an insight into the current developments in weed science. There are 17 chapters which deal with biology and ecology of weeds, weed management strategies, herbicides usage and modern concepts in weed science.

Chapters on biology and ecology of weeds deals with classification, propagation and dissemination, crop -weed association, allelopathy, soil seed bank and shift in weed flora in cropping systems.

Chapters on weed control familiarizes students on invasive alien weeds, concepts of weed prevention, control and eradication. Different methods of weed control, physical, cultural, chemical, biological and integrated weed management.

Chapters on herbicides enumerates the advantages and limitations of herbicide usage in India, herbicide classification, absorption translocation and selectivity of herbicides in plants, adjuvants and their uses, formulations, dosage and method of application, fate of herbicides in soil and compatibility of herbicides with other agro chemicals.

Chapter on aquatic weeds deals with problems due to aquatic weeds, classification, major weed flora in aquatic systems and their control.

Chapter on parasitic weeds gives an awareness on Orobanche, Striga, Cuscuta and Mistletoes, their life cycle and available control measures.

Chapter on weed management in field and horticultural crops contains information's on different techniques and chemicals adopted for the control of major weeds in field crops, plantations and orchard.

Chapter on modern approaches in weed science and managements deals with the use of biotechnological and molecular tools in weed management. We acknowledge the help from Dr. Parvathi M.S., Plant Physiologist, Kerala Agricultural University, in editing this chapter.

We hope this book will serve as a guide to students of weed science. Our experience both in the field and in the laboratories has been the guiding factor in the preparation of this book.. We express our gratitude to New India Publishing Agency for bringing out this publication

Authors

Contents

1

Introduction

India is basically an agrarian economy where 57.8% of rural households are engaged in agriculture (Situation Assessment Survey of Agricultural Household, 2021). 48.9 per cent of the Indian work force is employed in the agricultural sector, (ILO, 2016). The estimated Indian population for the year 2050 is 1.7 billion which will raise our food demand to ~400 million tonnes (Rao et al., 2018). In this context, Indian agricultural production system will have to take up the challenge to feed 17.5% of the global population with only 2.4% of land and 4% of the available water resources at its disposal. Currently India is the world's second largest producer of rice, wheat, and cotton, after China and the second largest producer of sugarcane, after Brazil. It is also the second largest producer of horticultural products in the world. However, productivity of these crops is much lower than in many countries and so there is a demand to enhance productivity to meet the demands of increasing population.

As per the information provided by the Directorate of Weed Research (DWR), Jabalpur India, weeds account for about one-third of the total losses caused by agricultural pests (DWR, 2015). Diverse climatic conditions in India favour the most adapted weeds to prevail and cause severe crop yield losses. Weed management in India is critical to improve crop productivity by minimizing crop yield losses and to alleviate other adverse effects of weeds in different ecosystems.

Weeds are considered as undesirable or valueless plants growing wild in a cultivated area. Jethro Tull first coined the term weed in 1931 in the book "Horse Hoeing Husbandry".They have been a problem ever since man started agriculture. However the term weed is a subjective one since many of these weeds have been utilized as source of food, fodder, herbal medicine fiber and fuel. Weeds form a major challenge to crop production. In the race between weed and crop for resources, ability of weeds to smother crops has always been a challenge for the farmer. Though herbicides have been a boon to keep weeds under check, environmental impact of these chemicals cannot be ignored.

Classical definition for weeds is the one given by Professor Beal, former weed scientist of Michigan state College as "A plant out of place". They interfere with human activities. Their negative characters outweigh their positive attributes. In many cases it can even be that their positive attributes have not been wholly understood. Hence Emerson(1878) defined weeds as plants whose virtues have not yet been discovered. According to Ross and Lembi, (1999) plants that are competitive, persistent, pernicious, and interfere negatively with human activity are defined as weeds. Weed refers to any plant that grows or reproduces aggressively or is invasive outside its native habitat.

To the farmer, weeds are a major problem in crop production due to economic concerns. They invade the crop field, compete with crop plants and finally take over the field adversely affecting crop growth and yield. It has been estimated that of the 250,000 species of recorded plants about 3% behave as weeds which is approximately 8000 species.

Major characteristics of these species are that they produce a large number of seeds, have very short life cycle, and devote most of their resources for seed production. Weed seeds are able to remain dormant for many years and retain viability in the soil for long periods of time. Popular saying is that one year of seeding leads to seven years of weeding. This is especially true for wild rice. Many species are propagated both vegetatively as rhizomes or stolon and as seed. Sedges belonging to Cyperacea family such as Nutgrass (*Cyperus rotundus*) and many perennial grass weeds eg. Alang alang (*Imperata cylindrical*) or Bermuda grass (*Cynodon dactylon*) can propagate both through vegetative means and through seeds. Many species have specialized structures such as awns, hooks, hairs or membranous structures which help in dispersal as they can cling to animals or humans and help in their spread.

Harmful Effects of Weeds

Domination of weed species in an ecosystem has a negative impact on agricultural output. In affluent countries loss from weeds is estimated to be around 5% while in less developed countries it is around 25%. However this can go upto 50 % depending on the crop. Field crops such as rice and wheat belong to this category. In developed countries large farm sizes help in adoption of mechanical and chemical methods of weed control while in developing countries, manual method of weed control is widely adopted. In India it is observed that loss from weeds is higher than that due to pest and diseases. All India Coordinated Research Project on Weed Management (ICAR) estimated yield and economic losses due to weeds in 14 major field crops based on data from 1581 On-Farm Research trials conducted in 18 states for the period from 2003 to 2014. It was observed that location (state), crop, and soil type explained the actual yield losses due to weeds in the farmers' fields. Actual economic losses were high in the case of rice (USD 4420 million) followed by wheat (USD 3376 million) and soybean (USD 1559 million).

Table 1: Estimated yield loss from weeds in different crops

S.No	Crop	% Yield loss
1	Soybean	31.4
2	Groundnut	35.8
3	Direct seeded rice	21.4
4	Maize	25.3
5	Wheat	18.6
6	Mustard	21.4
7	Sesamum	23.7
8	Green gram	30.8
9	Pearl millet	27.6
10	Transplanted rice	13.8

Reproduced from Gharde et al. (2018)

Compete for Food and Nutrients

Weeds reduce the quality and quantity of crop produce. They compete for resources such as water, nutrients, light, and space, lowering agricultural yields. It is reported that weeds absorb nutrients faster than crop plants and accumulate them in their tissues in relatively large amounts. Nitrogen is the most important element for which they compete but some species accumulate minor elements like sulphur, calcium or magnesium. This depends on the crop. Higher amounts of sulphur and magnesium was observed in *Cichorium intybus* and *Cirsium arvensis* than pasture crops (Harrington et al., 2006). In plantations weeds tend to smother young seedlings. Climbers like Mile a minute weed (*Mikania micrantha*) and Morning glory (*Ipomoea* sp.) are major problems in pineapple gardens, cashew and rubber plantations.

Tree parasitic weeds such as Loranthus (*Dendrophthoe falcata, Helicanthus elastica*) reduce productivity of fruit trees like mango and cashew. They also reduce wood quality of teak and mahogany.

Interfere with Cultural Operations

In many cases they interfere with cultural operations in the field and harvest process. They impede irrigation water flow and affect the free movement of man and machinery in the field. Weed control is vital to agriculture but it increase production cost.

Reduce Land Value

Weeds reduce land value. Perennial rhizomatous grasses such as Johnsons grass (*Sorghum halepense*), Bermuda grass (*Cynodon dactylon*), Wild sugarcane (*Saccharum spontaneum*) and many species belonging to Cyperaceae family make the land unsuitable for cultivation due to heavy infestation. Eradication of these species is extremely difficult since many of them are known to be fire resistant. Underground tubers or vegetative propagules require repeated tillage to exhaust reserve food materials from the tubers.

Parasitic weeds like Dodders (*Cuscuta* spp.), and Witchweed (*Striga asiatica*) are difficult to eradicate. They are also spread through seeds.

Affect Quality of Produce

They reduce quality of crop and fodder by contamination.eg presence of *Amaranthus* in fodder will adversely affect the forage quality. High nitrate content in *Amaranthus* can even be toxic to animals.

Weeds degrade quality of agricultural products that are sold. Contamination with weed seeds such as Thornapple (*Datura* sp.), Prickly poppies (*Argemone* sp.) and Wild mustard (*Brassica* sp.) is damaging to human health. Weed seeds in the produce can occasionally produce an unpleasant odour.

Host for Pest and Diseases

Weeds serve as alternate host for many disease organisms and also harbor insect pests. To reduce pest and disease infestation, field sanitation before planting is very important.

Table 2: Weeds that serve as alternate host for pest and diseases

Crop	Pest/ Disease	Alternate host
Rice	Stem borer	*Echinochloa, Panicum*
Rice	Blast	*Digitaria marginata*
Wheat	Black rust	*Agropyron repens*
Maize	Downey mildew	*Saccharum spontaneum*
Red gram	Gram caterpillar	*Amaranthus, Datura*
Ground nut	Leaf miner	*Borreria hispida*

Impair Health of Man and Animals

Weeds, like Parthenium produce toxic chemical and exhibit allergy in both animals and humans. They are responsible for chronic skin ailments such as dermatitis and asthma. Poison ivy (*Rhus toxicondendron*) Rag weed (*Ambrosia* spp.) Indian stinging nettle (*Tragia involucrata*) are some weed species that are skin irritants and cause itching and hay fever.

Impact of Aquatic Weeds

Aquatic weeds obstruct water flow in canals and drainage systems. This contributes to water pollution and de-oxygenation of the water bodies. The unhealthy environment is a threat to fishes and a breeding ground for mosquitoes. They interfere with water transport and navigation. They increase water loses through transpiration.

Benefits of Weeds

In spite of the many negative impacts of weeds, some plants usually thought of as weeds may actually provide some benefits. Many weed species possess certain characteristics or properties that allow them to thrive and multiply in undesirable location. The attributes of weeds that help to succeed in disturbed areas needs to be elucidated for trait transfer (Baker, 1974). Moreover they help to maintain the balance of nature.

Soil Binders

Many grasses are used as soil stabilisers to prevent soil erosion and loss of the fertile top soil. Weed species such as Bermuda grass (*Cynodon dactylon*) and Alang alang (*Imperata cylindrica*) are perennial grasses which quickly form a sod and help to prevent soil erosion in high rainfall areas (Holm et al., 1977).

They assist in the conservation of soil moisture and prevent erosion. Weeds help to preserve nutrients by reducing the quantity of bare soil exposed, especially nitrogen, which could be leached away, mostly from light soils. They also add organic matter to the soil.

Serve as Indicator Plants

Many weeds act as indicator plants which reflect on the field growing conditions, such as water levels, compaction, and pH. A high infestation of leguminous species such as Touch- me- not (*Mimosa pudica*) and Indian jointvetch (*Aeschynomene indica*) can be an indication of low nitrogen in the soil. Similarly Club rush (*Scirpus juncoides*) is seen in low phosphorus soils. Shaggy button weed (*Borreria hispida*) is an indicator of acidic soil condition while Bush morning glory (*Ipomea fistulosa*) is seen in saline or low lying soils likely to be under water during rainy season (Vidya, 2002)

Used for Human Consumption

In many countries weed species are used as vegetable. In Tamil Nadu, Black berry nightshade (*Solanum nigrum*) is considered as a vegetable with much medicinal value while in Europe it is a classified as weed. In Thailand Cruz and Price (2012) listed 43 weed species used as vegetables. They also found that weedy vegetables are an important resource for rice farmers in this region both as food and as medicinal plants.

Provide Feed and Fodder for Wild life

Some common grasses such as Torpedo grass(*Panicum repens*), Baryard grass (*Echinochloa crusgalli*) Crab grass (*Digitaria ciliaris*) and legumes such as Indian jointvetch (*Aeschynomene indica*) are used as fodder. Others such as Dandelions (*Taraxacam officinale*) considered as a menace in lawns are a source of nectar for pollinators like bees and beetles during early spring. Wild carrot (*Daucus carota*) flowering in field margins provides nectar and pollen for the adult phases of important natural enemies of vegetable pests.

Medicinal Value

Many herbs used in Ayurveda medicine are common weeds. Goat weed (*Scoparia dulcis*) is used for treating kidney stone (Murthy, 2005). Western stoneweed (*Lithospermum ruderale*) is used to control conception (Train et al 1941). Medicine prepared from Mistletoe (*Viscum album*) a tree parasitic weed is used in the treatment of cancer. Carry- Me –Seed (*Phyllanthus amarus*) is known for the treatment of liver diseases. Arrow leaf sida (*Sida rhombifolia*) for rheumatism and False daisy (*Eclipa alba*) is used for eye health and as hair tonic, Water hyssop (*Bacopa monnieri*) is used for gut infection and to enhance memory and cognitive functions.

Phytoremediation

Weed species are commonly used for phytoremediation.Many weed species absorb heavy metals from the soil and water and clean the ecosystem. *Eichhornia crassipes*, also known as lilac devil, is used in water purification. Kariba weed (*Salvinia molesta*) is reported to be a heavy accumulator of iron (Girija and Abraham, 2005).

Check Pest and Diseases

Many weed species serve as alternate hosts for the pests. Hence planting them at garden's perimeter helps to trap pests. Soybean loopers like to feed on sweet potatoes, tomatoes and also on weeds such as Oxalis, Kudzu (*Pueraria montana*) and Lantana. Cabbage loopers feed on weeds like Lambs quarters (*Chenopodium album*), Dandelions, and Curly dock (*Rumex crispus*).

Source of Organic Matter to Soil

Weeds are rich in organic matter. They are very fast growing species. Deep rooted weeds absorb nutrients from lower layers of the soil and bring them to the surface where they become easily available to crop plants. When they die and decompose, they add carbon and humus to the soil. This also helps in water retention. Many weed species are utilised for vermi composting. Incorporating weeds into the soil by tillage and other cultural practices enriches the soil organic matter content. Siam weed (*Chromolaena odorata*) and Lilac devil (*Eichhornia crassipes*) are commonly used for composting.

Genetic Reservoir

Many wild relatives of cultivated species may not be economically important but they are sources of many valuable genetic characters which impart stress resistance to the species and may be valuable in breeding programmes. Eg. *Sorghaum halepense, Saccharum* spp, *Oryza* spp. etc.

Provide Employment Opportunities

Many weed species such as Lantana, Eichhornia are used for making artifacts. Weeds belonging to cyperaceae family are used for making mats and baskets. Eco-friendly brooms are made from grasses. They provide employment for the local population. There are NGOs working with the local community to help them market these products.

Crop Weed Competition

When weeds can be a bane and boon to man, it is important to understand when and how to control the weeds in a crop ecosystem. As weeds are native to an ecosystem, they are much more adapted to the area and have a higher advantage than the introduced crop. There is a negative interaction between species which simultaneously demand the same resources. In the race for resource accumulation one suffers and one loses. Hence crop weed competition indicates competition between crop and weed in a natural ecosystem in response to resource utilisation for existence and superiority. As a rule weeds have higher growth rate and so during the initial growth period they tend to grow faster than the crop plant and compete effectively to smother the crop. Once the crop establishes itself it can out compete the weed species. The intensity of loss due to competition depends on

1. Time of infestation
2. Intensity of infestation
3. Type of weed and crop

Major resources that the weeds compete for include water, nutrients and light.

Competition for Nutrients

It has been observed that nutrient use efficiency of weeds is much higher than crop plants. Studies conducted by Dhima et al., (2006) indicate that most weeds take up nitrogen and potassium at a faster rate than crop plants. This was especially true for Wild mustard (*Sinapis arvensis*) in wheat crop. In many cases it has been observed that by managing the nitrogen application practices weed growth can be controlled. (Manoj, et al., 2020).

Competition for Water

Weeds grow rapidly and have an extensive root zone and more soil volume per plant. Hence they have more resource affinity and higher tolerance to climatic variations (Zimdahl,2018). They use more water than the crop plant. The water use efficiency that is the rate of water used in plant metabolism is higher for most weeds than the water lost through transpiration. Many of these features depends on the species, photosynthetic pathway, plant architecture, root length and distribution (Berger et al.,2015). It has been estimated that Lambsquarters (*Chenopodium album*) uses 550mm of moisture while wheat crop removes only 479 mm of water during a growth cycle. The weed is also able to absorb moisture from deeper layers of soil as compared to the crop due to its efficient root system.

Competition for Light

Light is an essential component for plant growth. Quality intensity and duration of light are the major factors that influence weed growth. Weeds seen as undergrowth in forest areas and plantations differ from those seen in open fields. Studies in pineapple plantations revealed that climbers like Bombay nawal (*Merremia vitifolia*) and Mile-a-minute weed (*Mikania micrantha*) which smother the crop can reduce the leaf area of pineapple and have a negative impact on productivity.

Managing Weeds

The implication of crop weed completion is not only on loss in productivity but also on farm level profitability. Climate change and environmental degradation have contributed to shift in weed flora. Use of chemicals in weed control has led to emergence of herbicide resistant weeds. Managing weeds for profitable crop production is a major challenge for the farmer.

References

Baker, H. 1974 . The evolution of weeds. Annu Rev Ecol Syst 5:1–24

Berger, S. T., Ferrell, J. A., Rowland, D. L., and Webster, T. M. 2015. Palmer amaranth (Amaranthus palmeri) competition for water in cotton. Weed Sci 63, 928–935. doi: 10.1614/WS-D-15-00062.1

Cousens, R. (1991). Aspects of the design ans interpretation of competition (interference) experiments. Weed Technology, 5(3), 664-673. http:// dx.doi.org/10.1017/S0890037X00027524.

Cousens, R., Brain, P., O'Donovan, J. T., & O'Sullivan, P. A. 1987. The use of biologically realistic equations to describe the effects of weed density and relative time of emergence on crop yield. Weed Science, 35(5), 720-725. http://dx.doi.org/10.1017/S0043174500060872

Dhima, K., Vasilakoglou, I., Eleftherohorinos, I., Lithourgidis, A., 2006. Allelopathic potential of winter cereals and their cover crop mulch effect on grass weed suppression and corn development. Crop Sci. 46, 345e352.

Djaman, K., and Irmak, S. 2012. Soil water extraction patterns and crop, irrigation, and evapotranspiration water use efficiency of maize under full and limited irrigation and rainfed settings. Trans ASABE 55, 1223–1238. doi: 10.13031/2013.42262

DWR. 2015.https://dwr.icar.gov.in/DWR%20Annual%20Report/2015-2016.pdf

Gharde, Y., Singh, P. K., Dubey, R. P., and Gupta, P. K. (2018). Assessment of yield and economic losses in agriculture due to weeds in India. Crop Prot 107, 12–18. doi: 10.1016/j.cropro.2018.01.007

Girija,T. and Abraham, C.T. 2005 Phytoremediation of mineral elements from water bodies using aquatic weeds. Indian J. Weed Sci.37 (1&2);155-156

Gisella S. Cruz-Garcia and Price L.L .2012. Weeds as Important Vegetables for FarmersActa Societatis Botanicorum Poloniae81(4):397-403

Harrington, K.C., Thatcher, A.A. and Kemp, P.D.2006. Mineral composition and nutritive value of some common pasture weeds. New Zealand Plant Protection 59:261-265

Holm, L.G., Plucknett, D.L., Pancha, J.V. and Herberyer, J.P. (1977). The World's Worst Weeds: Distribution and Biology. 609pp East-West Centre, United Press, Hawaii.

Manoj, K., Singh, R. P., Pandey, D. and Singh,G.2020. Nitrogen levels and weed control methods on yield and economics of wheat under zero-tillage conditions. Indian Journal of Weed Science 52(3): 241–244

Munger, P. H., Chandler, J. M., Cothren, J. T., and Hons, F. M. 1987. Soybean (Glycine max)–velvetleaf (Abutilon theophrasti) interspecific competition. Weed Sci 35, 647–653. doi: 10.1017/S0043174500060732

Murthy S.K.R. 2005. Vagbhata's Astanga Hridayam-3 Volumes.Krishnadas Ayurveda series.27 Pub.Chowkhamba Krishnadas academy,Varanasi

Naidu, S.G.R., Yaduraju,N.T. Gogoi,A.K (Eds).2005.Weeds that heal, National research, Centre for Weed Science, Jabalpur, India .pp120

Rao, A.N. Singh, R.G, Mahajan, G, and Wani, S.P. 2018. Weed research issues,challenges and opportunities in India. Crop protection.V (134):1-8.

Ross, S. M. , King, J. R. , O'Donovan, J. T. , Spaner, D. 2004. Forage potential on intercropping berseem clover with barley, oat, or triticale. Agronomy J., 96 (4): 1013-1021

Ross M. A and Lembi, C. A, 1999, Applied Weed Science 2nd ed, Prentice-Hall, Upper Saddle River, New Jersey. 452 p.

Singh, M., Kukal, M.S., Irmak, S. and Jhala, A.J. 2022. Water Use Characteristics of Weeds: A Global Review, Best Practices, and Future Directions. Front Plant Sci 12:794090. doi: 10.3389/fpls.2021.794090

Tilman, D. 1997. Mechanisms of plant competition. pp. 239-261 In M.J. Crawley (ed.) Plant Ecology. Blackwell Science, Oxford, London

Train,P.,Hendricks,J.R and Archer,W.A.1941. Medicinal uses of plants by Indian tribes of Nevada.PartIII. Flora of Nevada project. Issue by Division of Plant Exploration and Introduction, Bureau of plant Industry, United States Department of Agriculture, Washington, D.C..

Vidya, A.S. 2002.Weed dynamics in rice field: Influence of soil reaction and fertility. M.Sc (Ag). Thesis. College of Horticulture, Vellanikkara

Wychen, L. V. 2019. 2019 Survey of the Most Common and Troublesome Weeds in Broadleaf Crops,Fruits and Vegetables in the United States and Canada. Available online at: http://wssa.net/wp-content/uploads/2019-Weed-Survey_ Broadleaf-crops.xlsx. (accessed March 5, 2021).

Wychen, L. V. 2020. 2020 Survey of the Most Common and Troublesome Weeds in Grass Crops, Pasture and Turf in the United States and Canada. Available online at: http://wssa.net/wp-content/uploads/2020-Weed-Survey_Grass-crops.xlsx. (accessed March 5, 2021).

Zimdahl, R. L. 2018. Fundamentals of Weed Science. London: Academic Press.

Zimdahl, R. L. 2007. Weed-Crop Competition: A Review. Hoboken, NJ: John Wiley and Sons.

2

Classification, Propagation and Dissemination of Weeds

Weeds are diverse plant species of the ecosystem which can survive adverse condition and reproduce quickly. The global compendium of plants has listed 28000 species as weeds. According to Holm et al (1977) about 8000 weeds species have been reported from agricultural fields in 124 countries and of these, 250 species are considered to be problematic. It is also to be noted that weeds from 12 families constitute 68% of the world's worst weeds. Within these 12 families, weeds from poaceae, asteraceae and cyperaceae constitute 44% of the world's worst weeds. To develop effective measures of controlling weeds, it is important to understand the biology and propagation methods of these species.

Weeds can be classified in various ways- based on plant morphology, lifespan. habitat, association or origin. They are also classified as objectionable, noxious or poisonous weeds based on special considerations.

I. Classification Based on Morphology of the Species

Weeds are classified into 5 different classes as

1. Grasses: These are herbaceous plants with rounded or flattened stem in cross section. Leaves are narrow and attached in two rows along the stem. Leaf sheath will be split around the stem. Internodes are mostly hollow. In most cases ligules and auricles are present e.g., *Echinocloa crusgalli, Cynodon dactylon*
2. Sedges: Resemble grasses in appearance. Major difference is that ligules and auricles will not be seen. Leaf sheath will not be split around the stem. Stem in many cases will be triangular in cross section with solid internodes e,g., *Cyperus rotundus, Cyperus iria, Kyllinga nemoralis*
3. Broad leaved weeds: These are mostly dicotyledonous plants having broad leaves, with netted venation and exposed apical meristem. Stems will have true cambium with vascular bundles. e,g., Parthenium hysterophorus, *Eclipta alba, Mimosa pudica*

II. Classification Based on Life Span

This is based on ontogeny that is the time taken to complete its lifespan from seed to seed, accordingly, weeds are classified as

1. Annual- These are small herbs with shallow roots and weak stem. They complete their life cycle within a season or sometimes take upto one year. These weeds are normally propagated through seeds and in many cases, they are season bound e.g., *Ageratum conyzoides, Euphorbia hirta*
2. Biennial weeds take two years to complete their life cycle. During the first year or season the plants completes its vegetative growth. Flowering and seed setting takes place during the second year in concurrence with the season. Wild carrot, (*Daucus carota*), Biennial Wormwood (*Artemisia biennis*).
4. Perennial weeds live for more than 2 years. They are propagated both vegetatively and through seeds. Most of the perennial weeds have underground storage organs like rhizomes, stolon, bulbs or underground stem. Some are deep rooted. Eradication of perennial weeds is difficult. Unlike annuals, perennial weeds can regenerate from underground structures even when the top portion is destroyed. Perennial weeds are of three types
   a. Simple perennials- These are seed propagated species e.g., *Ipomoea carnea, Lantana camara.*
   b. Bulbous perennials- They are propagated by bulbs, bulblets as well as seeds e.g., *Allium vineale*
   c. Creeping perennials- They spread by the extension of stem laterally along the soil surface e.g., *Cynodon dactylon, Convolvulus arvensis*

Creeping perennials are further classified as

i) Shallow rooted perennials e.g., Hariyali grass(*Cynodon dactylon)* and Quack grass (*Agropyron repens*).

ii) Deep rooted perennials- Here the root system can go down upto 1 m or more e.g., Purple nut grass (*Cyperus rotundus*)

III. Classification Based on Habitat

Based on the place of occurrence weeds are classified into terrestrial weeds, aquatic weeds and aerial weeds

1. Terrestrial weeds are those which are commonly seen on cropped areas both upland and wet land, garden land, range lands and forest area. They can be seen in dry areas and areas where there is enough moisture for growth but do not prefer submerged condition. *Celosia argentea, Chenopodium album*

2. Aquatic weeds: These are plants seen in aquatic ecosystems. Plants growing in these situations are structurally modified. They can be grouped into four categories

   a. Submersed weeds: These are vascular plants, major growth phase occurs beneath the water surface. They have true roots, stem and leaves e.g.,*Ceratophyllum demersum* and *Hydrilla verticillata*

   b. Emersed weeds: They will be rooted to the bottom mud but the leaves and stem can be seen above the water level. In plants like *Nelumbo nucifera,* leaves may be broad or it can be like grass leaves as seen in the case of *Typha* spp. They are weeds which cannot rise and fall with the water level.

   c. Floating weeds: Common floating weeds can be seen as clusters like, Kariba weed (*Salvinia molesta*) or singly like water hyacinth (*Eichhornia crassipes*) or water cabbage (*Pistia statiotes*)

3. Aerial weeds: Epiphytic plants and parasitic weeds come under this category. Epiphytic plants such as Vanda may not have any adverse effect on the host plant. This also depends on how densely they cover the host plant. Parasitic species adversely affect the productivity of the host plant e.g. Mistletoes (*Dendrophthoe falcata*) and Dodder (*Cuscuta reflexa*)

IV. Based on Association

1. Season bound weeds: Many weed species exhibit seasonality of emergence. Summer weeds (Kharif annuals) generally germinate in summer or rainy season and complete their life cycle before winter season. These include *Tribulus terrestris, Echinochloa crusgalli.*

Winter weeds (Rabi annuals they germinate in winter season and complete their life cycle by spring or summer e.g., *Avena fatua*, *Phalaris minor*.

A study conducted on weed emergence in Kerala revealed that the weed species *Biophytum sensitivum* and *Mimosa pudica* dominated during the South West monsoon season while weeds such as *Curculigo orchioides* and *Desmodium gangeticum* were seen germinating during the North East monsoon period from September– October and were predominant in the field till February.

2. Crop bound weeds: Crop bound weeds are mostly parasitic weed species. They depend on the host partially or fully for their nourishment

a. Root parasites e.g. Broomrape (*Orobanche* spp.) are complete root parasites seen on tobacco and solanaceous crops. Witch weed (*Striga* spp) are partial root parasites on millets, sugarcane and upland rice

b. Stem parasites e.g. These can either be Holoparasites or complete parasites like Dodder (*Cuscuta* spp.) in lucerns and berseen or

Hemi parasites or partial parasites like Loranthus in (*Dendrophthoe falcata*) on fruit trees.

3. Crop associated weeds. Most of these are mimicry weeds which grow along with the crop. They require the same microclimate and habitat to survive. They are harvested along with the crop and contaminate the seed. In wheat, *Phalaris minor* and *Avena fatua* have been identified as associated weeds. In rice *Echinochloa crusgalli* and wild rice are some of the associated weeds

V. Based on Origin

1. Indigenous weeds: They are the native weeds of the country .e.g., *Phyllanthus niruri*, *Acalypha indica*, *Celosia argentea*

2. Alien weeds; introduced exotic weed species tend to become invasive in many cases. e.g., *Parthenium hysterophorus* was introduced from USA along with the imported wheat grain. *Salvinia molesta* was introduced from Brazil as ornamental plant. *Mikania micrantha* was introduced from North America due to its fast growth nature. During the second-world war it was grown for camouflaging troop movement. Many introduced weeds have become invasive and a major threat to many ecosystems throughout the world.

VI. Based on Special Consideration

Objectionable weeds: These are weed species very similar to the crop, mature at the same time and have similar height so that it is very difficult to separate seeds of the weed from that of crop, once they are mixed. They are also known as satellite weeds e.g., Wild rice (*Oryza spontanea*) and Barnyard grass (*Echinochloa* spp) in paddy. Little seed canary grass (*Phalaris minor*) in wheat

Noxious weeds: A plant is designated as a noxious weed by agricultural or other government agency when it is harmful to the environment, humans or livestock. Mostly they are introduced species which grow aggressively, multiply quickly without natural controls (native herbivores, soil chemistry, etc.), and display adverse effects through contact or ingestion. *Ambrosia psilostachya* which was introduced through wheat import was declared as a noxious weed by Karnataka government based on the National Invasive Weed Survey conducted from 2008- 2010. *Parthenium hysterophorus* also has been declared as a noxious weed by Karnataka government.

Poisonous weeds: Some plant species are poisonous to man and animals e.g.

Indian heliotrope (*Heliotropium indicum*) contains the alkaloids, tumorigenic pyrrolizidine which is toxic. Angels trumpet (*Brugmansia* sp.) all parts of plants in this genus contain the tropane alkaloids, scopolamine and atropine which is often fatal. Jimson weed (*Datura* sp.) contains tropane alkaloids, scopolamine, hyoscyamine and atropine. All parts of these plants are poisonous, especially the seeds and flowers. Ingestion causes abnormal thirst, hyperthermia severe delirium and incoherence, visual distortions, bizarre and possibly violent behaviour, memory loss, coma, and often death;

Reproduction and Propagation of Weed

Information on reproduction, propagation, life cycle, dissemination, growth and establishment of weed species will help to understand adaptive mechanisms that help weeds to succeed and persist in an environment.

Plants reproduce either sexually through seeds or asexually by vegetative means. Many weed species are propagated by both these method which accelerates their distribution and invasion. Prolific reproduction and efficient dissemination methods are key factors that account for wide spread and persistence.

Seed Propagation

Generally annual and biennial weeds are propagated through seed. Even seed propagated species become a problem because of this specific character of the weed seed.

1. High reproductive potential and longevity

    1. Seed production potential of many weeds are reported to be very high (Table 1).It has been reported that average arable soil can contain 30,000 to 3,50,000 weed seeds/m^2 and out of this only 2-10% of the weed seeds emerge each year from the soil seed bank (Walia et al., 2016)

Table 1: Seed production potential and longevity of major weeds seeds in soil

Weed species	Seed production / plant (Nos)	Longevity of seed in soil (years)
Barnyard grass (*Echinochloa crusgalli*)	7,00,000	5-7
Wild oats (*Avena fatua*)	750-2,000	4-6
Beggar-tick (*Bidens pilosa*)	7,000	3-5
Dodder (*Cuscuta* spp)	16,000	70
Lambsqaurters(*Chenopodium album*)	72,450	21-40
Common purslane (*Portulaca oleracea*)	52,309-18.00.000	20-40
Velvet leaf (*Abutilon theophrasti*)	48,000	15-40
Mile a minute (*Mikania micrantha*)	90,000 and 210,000 seeds/m^2	1-7

Source: Robert and Steven (2009)

Asexual reproduction occurs through vegetative parts like rhizome, stolon, bulbs, nuts, corm etc. Asexually propagated species are difficult to eradicate as they multiply and spread through fragments of stem, tubers or nuts along with agricultural implements, soil, farmyard manures or through planting materials etc. They persist in the soil for more than 5 years in some cases extending upto 80 years depending on the viability of weed seeds or their sexual parts.

Many such weed seeds do not lose viability even when they are ingested by animals. They still remain viable and are spread through cowdung or farmyard manure. Some such weed species include *Amarathus retroflexus, Chenopodium album, Portulaca oleracea, Echinochloa crusgalli* etc.

Table 2: Propagules of major weed species

Weed species	Vegetative propagule
Oxalis sp	Bulb
Cynodon dactylon	Stolon or Runners (above ground parts are known as runners)
Cyperus rotundus	Nut or tubers
Sorghum halepense	Rhizomes
Ambrosia psilostachya	Rhizomes
Thimothy grass (*Phleum pretense*)	Corm

Some of the worst weeds are propagated vegetatively and through seeds. *Cyperus rotundus* produces a large number of seeds but the main mode of propagation is through underground vegetative part, the tuber. According to Jha and Sen (1985), *Cyperus rotundus* has more than one type of underground stem. The tubers are rich in stored food materials with several viable buds. Their studies also indicate that desiccation reduced viability of the tubers. Tillage can bring the tubers to the surface, their prolonged exposure (Minimum of 10 days) would help to reduce weed infestation in an area.

Many of the vegetative propagules survive through winter and are common even in temperate regions. To understand the survival of seeds an experiment was stared in 1879 by William James Beal, professor of botany of Michigan Agricultural College. Dr. Beal's experiment is the oldest continuing experiment at one of the oldest colleges of agriculture. He buried 20 bottles with 50 seeds each mixed with sand of 20 different kinds of plants and buried them in the Beal Garden 20 inches below the soil with the mouths slanting downward so that water would not collect in them. He also left a map indicating their location for future scientists. He wanted them to dig up the seeds every five years and test for viability. The period was later extended to ten years by other scientists. In 1980, three species from the lot germinated. After 120 years, seeds of only one species, Moth mullein (*Verbascum blattaria*) still retained viability. It has also been reported that this species consistently germinated in all of the tests.

2. Seed Dormancy

Dormancy is an internal condition of the seed that impedes its germination even under favorable environmental conditions. Seeds can go into a state of dormancy immediately after attaining maturity which is often referred to as **primary dormancy** or innate dormancy possessed by seeds when they are dispersed from the mother plant.

Innate dormancy can be due to immature embryo, mechanically resistant seed coat, seed coat impermeable to water and gases or even due to presence of germination inhibitors.

Secondary dormancy refers to a dormant state that is either induced or enforced on non-dormant seed by un-favourable conditions for germination, mostly environmental conditions induce secondary dormancy. This regulates the seasonal germination of weed seeds.

Induced dormancy; Exposure of the ripe and mature seed to unfavourable conditions like dryness, high carbon dioxide concentration or high temperature or any other condition which is unsuitable for plant growth can put the seed to sleep again and it will continue even when the conditions change.

Enforced dormancy: This type of dormancy depends on environment interactions like lack of oxygen, water or air temperature which contributes to dormancy. In enforced dormancy seeds will germinate once the favourable condition returns.

Seeds under secondary dormancy can be encouraged to germinate by cultural operations like ploughing which brings the seeds to the soil surface. When such seeds are exposed to light and with irrigation they germinate. Farmers remove such weeds through stale seed bed technique. Induction of secondary dormancy in *Avena fatua* is attributed to the lack of oxygen (hypoxia).

Dormancy Inducing Factors

Light: Most of the annual weed species are sensitive to light. Since light penetrates only up to 1 to 2 mm depth depending on the soil type, even shallow burial can induce dormancy in such weed. Siam weed (*Chromolaena odorata*), *Portulaca oleracea* and *Amaranthus viridis* are photoblastic species which require light for germination.

Some weeds are negatively photoblastic which require darkness for germination e.g.s *Datura stramonium,* Hedge-mustard (*Sisymbrium officinale*)

Temperature: Every seed has a minimum, maximum and optimum temperature range for germination and any deviation from this range will induce dormancy in seeds. Alteration in temperature will break the dormancy and induce growth. According to Karssen (1982) seasonal periodicity in field-emergence of annuals is the combined result of seasonal periodicity in the field temperature and seasonal periodicity in the width of the temperature range suited for germination.

Oxygen

Oxygen is essential for germination. When there is deficiency of oxygen it not only inhibits germination but also stimulates anaerobic reaction that produce germination inhibitors. Oxygen can have both detrimental and beneficial effect on weed seeds. Many weed seeds fail to germinate under reduced oxygen (Corbineau and Come, 1995). It has been observed that the concentration of oxygen in the soil is around 19% (Benech-Arnold et al., 2000). Germination of *E. crusgalli* which is mostly a wet land weed commonly seen in rice fields, increased when oxygen concentrations in the soil was between 2.5 and 5% and declined when the oxygen concentration level was above 5%. In terrestrial weeds partial pressure of oxygen was observed to affect seed dormancy. Soil compaction reduces the partial pressure of oxygen which also reduces germination. This is the reason why seeds on soil surface germinate faster. Germination of *Melochia corchorifolia* was found to be accelerated with tillage and reduction of soil compaction. It prefers soil with sandy loam texture. Fruit of cocklebur (*Xanthium strumarium*) has two seeds of these the lower seed normally germinates in the spring when it reaches maturity while the top seeds remains dormant until the following year. Impermeability of the seed covering to oxygen has been shown to be the cause of dormancy in these seeds. However, in *Datura stramonium* hypoxia condition did not decrease germination (Benvenuti and Macchia.,1995).

Immature Embryo

In many plants seeds are shed before the growth of the embryo is complete. These species require an after ripening period before the seed can be ready for germination. e.g., *Typha latifolia* and *Polygonum* spp. In these species it is seen that embryos will not grow even when seed coat coverings are removed,

Impermeable seed coat: Hard and impermeable seed coat prevents the entry of water and gases into the seeds. Generally weeds belonging to the families Malvaceae, Leguminoseae, Chenopodiaceae and Solanaceae have hard seed coat. It also imparts mechanical resistance to germination and prevents the penetration of radical.

Germination inhibitors: Many endogenous inhibitors which are products of the secondary metabolic pathway such as coumarins, ferulic acid, abscisic acid accumulate in the seeds and prevent seed germination .A reduction in the concentration of these chemicals in the seed by leaching or other metabolic activities will induce germination.

After ripening requirement: In some cases development of the seed itself is not complete when they are shed from the mother plant. Such seeds require an after ripening period for the embryo development to be complete. This is especially true for plants like Ginkgo (*Ginkgo biloba*), European ash (*Fraxinus excelsior*), Holly (*Ilex aquifolium*), and many orchids where the embryo is disorganised and development is not complete.

Table 3: Factors inducing secondary dormancy in weeds

Inducing Factors	**Examples**
Anaerobic conditions	*Xanthium pennysylvanicum*
Darkness	*Latuca sativa, Phleum pretense*
Prolonged white light	*L.sativa, Nemophila insignis*
Prolonged far red light	*Arabis hirsute, Amaranthus caudatus*
Temperature above maximum	*Ambrosia trifida, Avena sativa*
Temperature below minimum	*Phacelia dubia, Torilis japonica*
Water stress	*Latuca sativa*

Source: Bewley & Black, 1994

Many seeds can remain dormant for many years It has been reported that seeds of legumes viz., *Indigofera hochstetteri, Indigofera oblongifolia* and *Tephrosia purpurea* can remain viable for 9,13 and 7 years respectively. (Williams,1979).

In many cases weed seeds will start germination immediately after maturation even if the seeds are not fully dried. Many species show staggered dormancy that is they may be viable immediately after maturity then they enter a phase of secondary dormancy and re- emerge after a specific period. Such behaviour has been noticed in weedy rice, which exhibits a secondary dormancy period of upto 10 months. These characters of weed seeds pose a challenge to their eradication.

Seed Dissemination

Weed seeds are efficient travellers. They have efficient dispersal mechanisms which take their progeny away from the parental plant to provide a non-competitive site for the young ones to grow. There are many natural resources which help the weeds to distribute their progenies beyond geographical boundaries.

Wind: Weeds have several adaptive mechanism that help them to travel by wind. Presence of wings, pappus, low seed weight, shape and structure of seed are some of them. Dispersal structures resemble parachutes, helicopters and gliders. These cottony coverings and parachute like structures help the seeds to float with the wind and get disseminated to far off places.

Pappus: This is seen in in asteraceae family. In species like *Tridax procumbens*, the calyx is modified into hairs which makes it a parachute like modification which help the seed to easily float with the wind.

Comose: In some weeds, we find that the seeds are covered with hair either partially or fully. Seen in members of asclepiadaceae like *Calotropis procera*. Here the mature and dry seed is attached to false septum which is detached by the wind currents. Seeds float in the air with the help of silky hair and are carried to distant places.

Baloon: Weed species like *Physalis minima* has modified papery calyx that encloses the fruit loosely along with entrapped air which imparts buoyancy to the seed and helps it to float in the air. In *Convolvulus microphyllus* seeds are enclosed in persistent calyx and corolla which also entraps some air. Here the seed along with these parts are taken by the wind.

Wings; Wings are modified appendages seen in species like *Acer marophyllum*.

Feathery persistent styles: In some weeds the fruit is an achene. When ripe, they have grey-white coloured, densely woolly styles, that allows them to blow away in the wind. The style in such cases is persistent and feathery e.g Timble weed or *Anemon cylindrica*, Indian globe thistle or *Sphaeranthus indicus*

Water: Water is a major agency for seed dispersal especially in the case of wet land weeds and aquatic species. Along with the surface water runoff the weed seeds reach the ditches and drains and through irrigation canals and river they spread to many parts. Many weed seeds remain viable for many years under water It has been reported that seeds of *Convolvulus arvensis* retain upto 50% viability when submerged for more than four years. Many seeds have adaptation that help them to remain buoyant in water . Fruits of *Rumex crispus* and *Sagittaria* spp. have corky wings that help them to float.

Human beings: Intercontinental weed dispersal is mainly through human interaction. Many notorious weeds are spread through import of agricultural produce. *Parthenium hysteophorus* came to India through shipment of wheat imported from USA in 1960. Once such species are introduced further spread can be through wind, water or animals. Lailac devil or *Eichhornia crassipes* has spread to different parts of the world by human introduction as an ornamental plant. Today it is one of the most troublesome weed in aquatic system.

Animals: Seed dispersal by bird and animals is the most effective and common method of dispersal for most of the major weed species, It is also known as Ornithochory and Mammaliochory respectively. Weed seeds have special appendages such as barbs, hooks, spines and rasps that cling to the fur of the animals and help in long distance dispersal. This is especially true in case of migratory birds and animals.

Presence of long awns: Many grasses have stiff awns or spiked glumes which adhere to the mammals and birds. Some weed species like *Cenchrus biflorus* have pointed and stiff projections which are broad at the base and pointed at the apex while seeds of *Cenchrus sentigerus* possess short hairy awns. These entangle with the hairs of the animals and get dispersed.

Presence of sharp spines: Some weed seeds have elongated terminal spines. The spiny tipped bracts and bracteoles attach to the cloths of humans and the skin of animals and fruits get dispersed e.g. *Tribulus terrestris*

Adhesive fruits and seeds; Weeds have mucilaginous gelatinous fruits. These adhere to the animals when they feed on them. Most of the tree parasitic weeds such as *Dendrophthoe falcata, Helicanthus elastica* have seeds which are edible to birds and squirrels . The seeds will stick to their beak or skin and thus they spread from tree to tree and plantation to plantation.

Fleshy fruits Weeds with fleshy fruits are consumed by animals and birds . The seeds are protected so they pass through the alimentary canal of animals and get spread through faeces e.g *Withania somnifera, Solanum nigrum, Lantana camara*

It is observed that ants carry a lot of weed seeds and store them in their holes. They collect seeds of *Dactyloctenium aegyptium*, *Cenchrus sentigerus* etc. This is also known as Myrmecochory. The distance travelled in such cases may not be too far.

Farm Machinery: Tillage and harvesting equipment's disperse weed seeds and vegetative propagules. Cultivation equipment's may drag rhizomes and stolons dropping them to start new infestation. Combine harvesters and threshing equipment's collect seeds of mimicry weeds along with the crop seed and get involved in their distribution to different areas as they move from field to field.

Seeds: Seeds are major sources of weed spread especially in the case of annuals. Testing agencies clearly state the level of contaminants in the seed lot but even such low levels if left untendered in the field can cause a lot of damage and eventual spread of weed in that locality. It is estimated that a single *Cuscuta* plant can spread to occupy approximately 25m^2 during a single season so just .001% contamination with *Cuscuta* seeds is sufficient to completely infest a legume crop field.

Manures: Today people prefer organic substances like farm yard manure, vermi compost, poultry manure, cow dung etc for cultivation. In most cases optimum temperature conditions are not attained in the composting pits. Hence, most of the weed seeds remain viable even after composting. Places where organic manure is used there will be a rich diversity of weed species.

References

Abraham, M., and C.T. Abraham. 2005. Biology of mile-a-minute weed (Mikania micrantha HBK), an alien invasive weed in Kerala. Indian J. Weed Sci. 37:153 – 154

Benech-Arnold, R. L., Sánchez, R. A., Forcella, F., Kruk, B. C., and Ghersa, C. M. (2000). Environmental control of dormancy in weed seed soil banks. Field Crops Res. 67, 105–122. doi: 10.1016/S0378-4290(00)00087-3

Benvenuti, S., and Macchia, M. 1995. Effect of hypoxia on buried weed seed germination.Weed Res. 35, 343–351. doi: 10.1111/j.1365-3180.1995.tb01629.x

Bewley J.D., and Black, M., (eds.). (1982). "Viability and longevity," in Physiology and Biochemistry of Seeds in Relation to Germination, (Berlin: Springer), 1–59. doi: 10.1007/978-3-642-68643

Brooks, S. J., F. D. Panetta, and K. E. Galway. 2008. Progress towards the eradicationof mikaniavine (Mikania micrantha) and limnocharis (Limnocharis flava) in northern Australia.

Cara Giaimo.2016. The World's Longest-Running Experiment Is Buried in a Secret Spot in Michigan. Natureculture cara@atlasobscura.com

Corbineau, F., and Côme, D. 1988. Primary and secondary dormancies in Oldenlandia corymbosa L. seed. Life SciAdv Plant Physiol. 7, 35–39.

Corbineau, F., and Côme, D. 1995. "Control of seed germination and dormancy by the gaseous environment," in Seed Development and Germination, eds J. Kigel, and A. Galili (New York, NY: Marcel Dekker), 397–424. doi: 10.1201/9780203740071-14

Day M D,. Clements D R, Gile Christine, . Senaratne W. K. A. D, Shen S, Weston, L A. and Zhang F2016. Biology and Impacts of Pacific Islands Invasive Species. 13.Mikania micrantha Kunth (Asteraceae) Pacific Science, 70(3):257-285 https://doi.org/10.2984/70.3.1

Holm, L.G., Plucknett,D.L., Pancha,J.V. and Herberyer,J.P. (1977).The World's Worst Weeds: Distribution and Biology. 609pp East-West Centre,United Press, Hawaii.

Jha P. K. and Sen D.N. 1985. Adaption and Survival Strategies in Cyperus rotundus L.: A Perennial Weed of Pantropical Distribution. J. of Folia Geobotanica & Phytotaxonomica. Vol. 20, No. 3 (1985), pp. 281-289

Karssen, C. M. 1982. "Seasonal patterns of dormancy in weed seeds," in The Physiology and Biochemistry of Seed Development, Dormancy and Germination, ed A. Khan (Amsterdam: Elsevier Biomedical Press, 243–270.

Robert, G. H. and Steven, S.J. 2009. Weed Seed Banks: Biology and Management. J. of .Prairie soils and crops scientific: perspectives for innovative management. (2) 1916-7091

Sathyanarayana, N., Kulkarni, N., Prasad, T., Sanjay, M. T. and Satyagopal K.2014. Specific Surveillance and Delimiting Survey of Ambrosia psilostachya in Tumkur District of Karnataka, India. Indian Journal of Plant Protection.Vol.42.78-82

Sindhu,P.V.,Thomas,C.G. and Abraham, C.T. 2010. Seed bed manipulations for weed management in wet seeded rice. Indian J Weed Sci. 42, 173–179.

Walia. U.S., Bhullar, M. S. and Singh, M. 2016. Weed biology and ecology. In Yaduraju, N. T. Sharma, A. R. and Das, T. K. (Eds.).2016. Weed Science and Management.402p.Indian Society of Weed Science, Jabalpur and Indian Society of Agronomy, New Delhi.

Williams, RD. 1979. Photo periodic effect on reproduction biology of purplenut sedge (Cyperus rotundus) Weed Science 26 ;539-42.

Zhang, F., Li, T., Xu, G., Wu, D. andZhang, Y. 2011. "Comparative analysis of growth types and reproductive characteris-tics of Mikania micrantha." [In Chinese.] Chin. Bull. Bot. 4659-66.

3

Survival Strategies and Weed Persistence

Weeds have specific characters which helps them to adapt to any stressful situation and successfully, compete with the crop. They have survival strategies which help them to overcome vagaries of nature like extreme cold, heat, drought, biotic stresses and soil abnormalities. Their most important characteristic is the ability to succeed in any adverse situation even in disturbed soils. The age old war between man and weeds makes it pertinent to understand the characteristics of weeds which make them better competitors in the field. With the advent of genetic engineering it is currently possible to identify and transmit useful traits to crop plants. Hardiness and adaptive mechanisms of weed species have to be studied to understand their success in adverse conditions.

Characteristics of Weedy Plants

1. Growth Strategies

Some common characteristics of weedy species are aggressive growth, competition with other plants for light, water, nutrients, and space, ability to grow in a wide range of soils and adverse conditions, and resistance to control measures. Disturbances in vegetation cover and changes in environmental conditions due to natural events or human activities and management practices create opportunities for species with adapted life cycles and growth characteristics to become established, reproduce, and colonize a site. Ability to adapt quickly to new environmental conditions and survive under disturbed and stressful situation indicate potential for invasive growth. Combinations of these characteristics are commonly exhibited in plant belonging to families, Asteraceae (aster), Brassicaceae (mustard), Polygonaceae (knotweed), Fabaceae (pea), and Euphorbiaceae (spurge), as well as others.

Rapid growth rate is exhibited by some ruderals (plants growing on waste land or rubbish) which are mostly mosses such as *Funaria hygrometica* commonly seen in woods or recently burnt lands and annual meadow grass (*Poa annua*). These species exhibit stress tolerance and are capable of exploiting unproductive environments or niches. They grow under high stress and low disturbance

Competitive Weed Species have the Following Growth Characters

a. Tall stature: Height is an important character that allows plants to compete for light and space. Most weed species grow taller than the crop plant e.g Baryard grass (*Echinochloa crusgalli*) and Weedy rice (*Oryza spontanea*) in rice.

b. Plant. characters that allow an intensive exploitation of the environment above and below the ground: An extensive root system such as a densely branched rhizome as in Stinging nettle Urtica dioica and Creeping soft-grass Holcus mollis, or an expanded tussock structure as in tufted Hair-grass *Des-champsia cespitosa.*

c. High relative growth rate: As compared to crop plants, weeds have higher relative growth rate which enhances their competitiveness.

Higher Resource Partitioning for Seed

Many weed species are ephemerals that have very short life cycle. They grow and reproduce quickly and generally produce more than one generation in a year growing only during favourable period e.g. *Echinochloa colona*, *Isachne miliacea*. These plants can adapt to harsh and drastic conditions like the desert or a shot spring as in the Tundra (arctic region).

2. Mimicry with Crops

In biology mimicry may be defined as a phenomenon that is characterised by the superficial resemblance of two or more organisms that are not closely relat-ed taxonomically. Vavilovian mimicry, also known as weed mimicry or crop mimicry is the type of mimicry where a weed evolves to share one or more characteristics with a domesticated plant species through generations of artificial selection. The Russian plant scientist, Nikolai Vavilov observed that cer-tain weeds adapt to weed control practices to survive in pre historical agrarian societies. This is especially true in the current context when agricultural machineries may contribute to selection and propagation of mimicry weeds. Most of these species have a synchronized growth with crop plants.

Table 1: Mimicry weeds

Crop	Mimicry weeds
Flax (*Linum usitatissimum*)	False Flax (*Camelina sativa*),
Sesamum (*Sesamum indicum*)	Winged -seed-sesamum (*Sesamum alatum*)
Oats (*Avena sativa*)	Wild oats (*Avena fatua* and *Avena sterillis*)
Rice (*Oryza sativa*)	Barnyard grass *Echinochloa crusgalli,* Weedy rice (*Oryza spontanea*)
Wheat (*Triticum aestivum*)	Canary grass (*Phalaris minor*)

3. Plasticity of Weed Growth

Weeds have the ability for rapid phenotypic adjustment to environmental changes which is known as phenotypic plasticity. It is defined as the ability of an organism to change its morphological and /or physiological features after exposure to different environmental conditions. Though most plant species exhibit this behaviour, the range of plasticity is very high in weed species. It is this feature which helps the weeds to improve their performance under new set of conditions. It acts as a buffer against extinctions and also helps in ecological adaptation and helps evolution. It improves the chances of survival and helps reproduction and colonization of weed species. The is evident in many common weed species. *Isachne miliacea* is a common weed of rice fields the ability of the weed to change its growth characters in different soil types according to the environment and availability of resources is indicated in Table 2

Table 2: Variation in phenophases(days) of *Isachne miliacea* under different soil types

Phenophases	Sandy loam	Clay loam	Acid saline	Lateritic
Seed germination	8-10	8-12	5-8	12-15
Tillering	15-19	18-21	13-17	20-25
Flowering	36-52	38-55	28-60	41-57
Seed formation	40-62	42-65	33-68	50-62
Seed maturation	54-70	55-68	40-74	56-70

Source: Suada (2015)

Other phenotypic characters that can be modified according to resources are the following

- Plant height
- Number of shoots
- Internodal length
- Partitioning
- Leaf Area Ratio (LAR)

- Root Length Ratio (RLR)
- Nutrient Use Efficiency (NUE)
- Specific Leaf Area (SLA)
- LMR(Leaf Mass Ratio)
- Leaf angle
- Total thickness of leaf
- Leaf density
- Stomatal density

Modification in these parameters according to the surroundings helps the weed species to successfully compete with the crop plants.

4. Prolific Seed Production

Ability to produce large number of seeds within a very short period is a character seen in weed species. Within a crop season, major weed species are capable of completing one or two cycles of growth. In addition to this some are capable of vegetative reproduction which is a major hurdle in weed eradication. Weed seeds can remain viable in the soil seed bank for many years. This helps the weeds to infiltrate undisturbed regions and proliferate if the site becomes disrupted in future.

Table 3: Seed production potential of common weed species

Weed Species	Average no. of seed produced/plant
Amaranthu spp	1,96,000
Leucas aspera	5000
Tridax procumbens	500-1500
Cynodon dactylon	170
Cyperus rotundus	40
Commelina benghalensis	2,450
Datura stramonium	13,900
Celosia argentea	11,812
Trianthema sp	52,000
Solanum nigrum	1,78,000
Eleusine indica	41,200
Mimosa invisa	72,500

Source: Steven (1932.1957) Jayasree (2005) Robert and Steven (2009)

5. Seed Longevity

Longevity is the life span of the weed seed. Longevity in the soil depends on many factors like dormancy, soil, environment, predation etc. Major reason for producing a large number of seed is to overcome these threats in the soil seed bank. Seed longevity is an important attributes that enhances survival and spread of weed species.

Table 4: Longevity of weed seeds buried in soil

Weed species	Longivity in soil (Years)
Covolvulus arvensis	20+
Sorghum halepense	20
Cirsium arvense	21
Datura stramonium	40
Amaranthus viridis	40

Source: Holm et.al (1977); Martin and Burnside (1980); Kingman and Ashton (1982); Salisbury and Rose (1992)

6. Change in Photosynthetic Pathways

All green plants fix CO_2 by the process of photosynthesis. This process is mediated by the enzyme RuBisCo. Depending on the environmental conditions, plants have evolved three photosynthetic pathways for CO_2 fixation. In C3 pathway CO_2 fixation is in the Calvin cycle. Major drawback of this system is that the carbon fixing enzymes of Calvin cycle will like to grab O_2 rather than CO_2 and release CO_2 by the process of photorespiration. C4 pathway uses a modified Calvin cycle. Here they prevent the oxygenation of RuBisCo from atmospheric oxygen by fixing CO_2 in the mesophyll cells and later the oxaloacetate and malate formed is ferried to RuBisCo and rest of the Calvin cycle enzymes are isolated in the bundle sheath cells. Photorespiration is absent in C4 plants. PEP Carboxylase which is the first acceptor of CO_2 and the products, oxalate and malate formed in this process are 4 carbon compounds so it is known as the C4 pathway. Plants adapted to dry environments, use the CAM (Crassulacean acid metabolism) pathway to minimize photorespiration. In CAM plants process of CO_2 fixation and assimilation are separated by time. Plants adapted to dry situations such as pineapple and cacti use this mechanism to minimize photorespiration. During the night they open the stomata and take in CO_2 which is fixed into organic acids like malate or oxalate and store it in the vacuoles. During day time the malate is transported out of the vacuole and is broken down to release CO_2 which enters the Calvin cycle. So the process of photosynthesis is carried out during day time by keeping the stomata closed. This helps to prevent transpiration and photorespiration.

It has been observed that many of the world's worst weeds have C4 photosynthetic pathway. Moreover some species can shift the photosynthetic pathway under conditions of stress. This has been reported in Ice plant (*Mesembryanthemum crystallinum*) which is basically a C3 species under normal condition. When it is subjected to salt or water stress it shifts to CAM pathway and keeps the stomata closed. The CAM pathway in plants is rapidly and reversibly induced in response to temperature, high light (Haag□Kerwer et al., 1992) and drought stress (Borland et al., 1992; de Mattos and Lüttge, 2001).

Table 5: Photosynthetic pathways of major weed species

Weed species	Photosynthetic pathway	Number of countries where the weed is a problem
Cyperus rotundus	C4	91
Cynodon dactylon	C4	90
Eleusine indica	C4	64
Echinochloa colona	C4	87
Echinochloa crusgalli	C4	65
Sorghum halepense	C4	51
Portulaca oleracea	C4	78
Chenopodium album	C3	58
Eichhornia crassipes	C3	50

7. Self- regeneration

Weeds are self-sown and weed seeds are carefully stored by nature in soil seed banks where they remain viable for a long period. They do not require specific cultural practices for germination. They depend on seasonal cues for germination and establishment. Vegetative multiplication also occurs spontaneously in perennial weed species. Such plants become aggressive and compete with their neighbours. Many successful weed species propagate both through seeds and vegetative parts.

8. Weed Succession

Most of the weed species can cross breed leading to the development of new phenotypes with wider adaptability. This is observed in field crops like rice wheat etc. In the case of wild rice it is observed that they easily cross with the rice cultivars. Progenies from such crosses show wide variation. Wild rice is normally tall with awns on the seeds. Progenies developed seem to be shorter than the cultivated rice variety and in many cases without awns. Early shattering character is another major trait of wild rice. This is carried over to the next generation in most cases. These characters of the weed make their eradication difficult.

9. Resistance to Herbicide

Herbicides revolutionized weed control as they provide the most reliable and economic form of weed management. Evolution of weeds with herbicide resistance is a challenge faced by farmers. Weed species resistant to chemical herbicides have specific mechanisms to detoxify chemicals. There can be both anatomical and physiological features of crop plants which impart resistance to herbicides. Continuous use of the same herbicide in a locality would impart resistance to weed population. This results from selection of individuals in weed population that undergo rare mutations that enable them to survive, reproduce and eventually predominate after repeated application of the same chemical. Currently 459 cases of herbicide resistant weeds have been reported globally. This includes 246 species, out of which 143 are dicot weeds and 103 monocot weeds. Herbicide resistant weeds are a major problem in field crops such as wheat, corn, soybean, rice, cotton and barley. In India it was first reported in wheat crop from Hisar during1992-93 in *Phalaris minor*, a troublesome mimicry weed in wheat crop. Continuous use of isoproturon in wheat crop led to the emergence of herbicide resistant species in *Phalaris minor* (Malik and Singh, 1993, 1995).

Crop Weed Competition

In nature there is a constant war between communities for various growth resources such as nutrients, light, water, space, CO_2 and O_2. Success of any species depends on its capacity to capture resources and effectively utilise them for growth.

According to the Oxford dictionary "Competition is the action of endeavouring to gain what another endeavours to gain at the same time, the striving of the two or more for the same object – rivalry".

According to Grime's Theory "competition is the tendency for neighbouring plants to utilize the same resources. The success in competition however is largely dependent on the capacity of species for resource capture. The competitive ability of plants/weeds depends upon their aggressiveness in acquisition of growth factors and utilizing them in rapid tissue build-up for growth"

Thilman (1982) defined Competitive success as the ability of plants to draw resources down to low level and to tolerate those low levels. It depends highly on their persistence and continued growth, utilizing growth factors minimally. According to him a plant which showed rapid tissue build up with the lowest intake of resources is a good competitor.

Weeds often appear more adapted to an ecosystem as they are the native flora of the region. Hence they have a tendency to aggressively compete with the crop plants for resources. According to Zimdahl (2007) "crop yield reductions is in proportion to the amount of light, water or nutrients weeds use at the expense of a crop". Generally it is observed that increase in 1kg of weed growth will decrease 1 kg of crop growth. Competition is mainly for natural resources. Plasticity helps the plant to navigate away from the site of competition and help in resource capture. Competition induced plastic responses in both shoot and root architecture is an important aspect of competitiveness of a species

Mineral Nutrients

Among the nutrients, nitrogen is the most important element for which crop and weeds compete. Weeds usually absorb mineral elements much faster than many other crop plants and accumulate them in their tissues in relatively large amounts. Intensity of competition depends on density and type of weed flora and period of emergence of the weed in relation to the crop (Malik & Singh, 1993)

Studies indicate that *Amaranthus* spp. accumulates over 3% more of nitrogen in dry matter as compared to crop plants. Harrington et al (2006) observed higher levels of sulphur and magnesium in the weed species, *Circium arvense* and *Cichorium intybus* as compared to pasture crops.

Moisture

Weeds compete with crop plants for water. They have extensive root system which allows faster absorption of water from the soil. Normally a field with weeds will show faster symptoms of wilting than when it is kept in a weed free condition. In rain fed crop there will be a constant competition for water between the crop and weeds. Immediately with the receipt of rain, weeds germinate faster and try to dominate the ecosystem. A comparison of the root system of rice and *Echinochloa* sp. shows that rice roots easily become dysfunctional due to deposition of iron oxide or other substance but roots of *Echinochloa* sp., a major weed of rice ecosystem, do not show such symptoms and seem healthier than the rice roots.

Light

When moisture and nutrients are plentiful light becomes the most important factor that reduces yield. Light is one of the most important constituent of photosynthesis. Leaf shading induces competition for light. It occurs when weeds have an edge over the crop with respect to height. Then leaves of weeds can block light falling on the crop. This starts during early growth phase when a weed grows very fast and completely smothers the crop. High intensity of *Phalaris minor* in field will completely smother wheat crop. Phenotypic plasticity is observed in rice field when the population of the weed *Echinochloa crusgalli* is higher than the threshold value of 2 weeds/sq.m. At this level it does not compete with rice plant but when the population of the weed is higher than the threshold level, it tends to grow taller than the rice crop and shading is a major cause for yield reduction. Available light is observed to be a major factor in the competition of *Cyperus esculentus* with maize crop. Here weeds are seen between crop rows, where light reaches the soil surface and not within crop rows

Shetty et al. (1982) was of the opinion that dicots are less sensitive to light than monocots, as dicots usually germinate after the monocots, in many cases even after establishment of the crop.

Space

Plant to plant spacing of a crop is a major factor that decides growth character of plants. Plants grown under wider spacing have wide spread branches and a prolific root system for resource accumulation and also there is sufficient light infiltration through canopy to soil. All these factors contribute to weed growth. In many crops like turmeric, potato and sugarcane it is seen that as the crop takes longer time for growth and establishment weeds grow without any competition. Spacing trials conducted in rice show that higher yield is obtained with closer spacing and weed counts is also much lower.

Major Factors that Influence Crop Weed Competition Include the following

a. Density of the weed

As the weed density increases it diverts the resources from the crop to the weed. Weed density is normally higher in disturbed agricultural soils. In many field crops, low population of weeds throughout the cropping season is not always negatively competitive to the crop but when the population goes beyond the threshold level, competition for available resources required for weed growth affects growth of the crop and productivity.

b. Crop density

Crop density is also an important factor that affects competition. Lower spacing increases the number plants of the crop in one sq.m area. This reduces the infiltration of light to soil. Since most weed species require light for germination it helps to check weed growth.

Type of weed species

Weeds also differ in their ability to compete with crop at similar density levels. Some weed species have allelopathy which affects the germination and growth of crop plants. Others have an extended root system that helps in faster resource accumulation.

c. Type of crop species and their varieties

Crops differ in their ability to compete with weeds. Extended periods of weed competition, especially during major growth phases like tillering, panicle initiation etc will adversely affect crop productivity. Duration of the crop varieties is an important character that decides the extent and time of weed free phase in the crop cycle. Some crop species have high competing ability. Among the cereals wheat, maize and barley, barley crop is observed to be a better competitor than the others. This is mainly attributed to the extensive root system which develops within the initial three weeks of barley crop. Sesamum crop showed lower competitive ability with the weed *Melochia corcorifolia* (Sreepriya et al. 2018). Rice varieties with competitive ability was screened at RARS Pattambi and it was observed that shade tolerant varieties like "chennellu" performed better under upland conditions. This was mainly due to their higher competitive ability and shade tolerance (Girija, et al. 2002)

d. Soil factors

Growth characters of many weed species depend on the soil type e.g., growth of *Isachne miliacea* is faster with prolific seed production in soils with high organic content like Kuttanad and Kole lands while the same weed has a lanky growth with lesser seed production in lateritic soils. Increasing soil fertility was found to enhance weed growth. Studies indicate fertiliser application improved both crop yield and also growth of weeds. Benefit of fertiliser application was higher in a weed free crop. In the field it is important to remove weeds before fertiliser application.

e. Time of germination

Timing of weed germination is found to be the most important factor that decides yield loss. This is because the first plant that gets established in an area gets a distinct competitive advantage over other plants. Early season weed competition will reduce yield more than the late season weed competition

Consequences of Crop Weed Competition

When crop and weeds compete for the same resources in a cropped ecosystem one of the competitor suffers a setback in growth. This is observed as nutritional deficiency symptom, reduced leaf area development, lower yield attributes, lower uptake of water and other inputs by crop. These factors lead to lower total dry matter production and affects proper development of yield attributes. This in turn reduces yield. The incidence of pest and diseases also increase.

Critical Period of Crop Weed Competition

Plants differ in their growth characters. In case of crop plants, growth is slower during the early stages. Some plants experience transplanting shock which inhibits the early growth rate of plants. Weeds are generally native floras so they grow very fast and compete with the crop plants for water, nutrients and light. Critical period of weed competition is the period between early growth during which weeds can grow without affecting crop yield and the point after which weed growth does not affect the yield. It is also defined as the short time span in the ontogeny of crop growth when weeding will result in the highest economic returns. Initial growth phase of crop is generally the most sensitive period. In most cultivated field crops, it comes to one third of the crop duration. There will be maximum yield loss if weeds are not removed during this period. To avoid yield loss it is important to control the weeds during this period which varies from crop to crop. Cousens and Mortimer (1995) suggested that the critical period can be calculated using regression analysis. Table gives the critical period of crop weed competition for some major field crops.

Table 6: Critical period of crop weed competition for some crops

Crop	CPCWC (Days after planting)
Rice (direct eeded)	15 to 45
Rice (transplanted)	30to45
Dwarf wheat	30 to 45
Tall wheat	35 to 50
Chick pea	30 to 60
Maize	30 to 60
Sorghum	15 to 45
Sugarcane	30 to 120
Greengram	15 to 30
Onion	30 to 75
Sesamum	20-30

Thresholds Level

It has been observed in the field that weed competition is not always negative. Based on the plasticity of plant growth, competition activates the growth of the crop. Weed control operations are also costly. So it is important to identify the density of weeds at which the cost of control equals the benefits obtained from the crop yield. This is known as the threshold value. In a rice field it was estimated that presence of *Sacciolepis interrupta* more than 2 weeds per sq m will adversely affect the yield of the crop where as the presence of 32 seedlings of *Isachne miliacea* was not so damaging. So in such situations, weed control operations may be adopted only when the population of the weed is above this level.(Shiji, 2001)

Weed Shift

When a change occurs in the composition of species in a community of weeds as a result of environmental or human made changes in an agricultural system it is referred to as weed shift. This is not a new phenomenon it has been happening in nature through the eons, we are concerned when some biotypes of weeds are susceptible to certain management practices while others are not affected by these management practices. This contributes to a change in the weed population dynamics of the area. Weed floral shifts can be observed in field crops. In rice field, Barnyard grass (*Echinochloa* sp.)is found to be replaced by Chinese sprangletop (*Leptochloa chinensis*) and *Marselia quadrifolia* (Yaduraju and Kathirasen, 2003).

This has been reportedfrom Cauvery delta and in kole lands of Kerala. In areas where 2,4-D is used frequently there is a shift to grasses. Selection of herbicides and cultural practices adopted in cultivation influence weed shifts. Continuous use of pre-emergence atrazine in maize has resulted in the shift of flora to *Commelina banghalensis*, *Brachiaria ramosa* and *Ageratum conizoides* in Utter Pradesh. Adoption of zero tillage in the eastern Indo-Gangetic Plains has resulted in significant increase of perennial weeds such as Purple nutsedge (*Cyperus rotundus*) and Bermuda grass (*Cynodon dactylon*) (Malik and Kumar, 2014).

Changes in weed flora in any ecosystem should be periodically monitored to asses emerging weed problems and to devise effective weed control measures.

Soil Seed Bank

Soil acts as a major reserve of weed seeds and vegetative propagules. It contains both new seeds as well as seeds that are accumulated from different seasons. It is a dynamic system where in both seed deposits as well as seed withdrawal takes place from season to season and year to year.

Weed seeds reach the soil through various dispersal mechanisms as well as through seed rain. Once they reach the soil they become part of the soil seed bank. Some enter a period of dormancy this mechanism helps weed species to be distributed to unpredictable environment. Some seeds are buried deeply in the soil and enter a phase of secondary dormancy. Non dormant seeds germinate and die prematurely if conditions are not favourable. Seeds are also lost by predation.

There are different types of soil seed banks. These include transient weed seed bank and persistent weed seed bank

Transient seed bank: In this case seeds remain viable in the soil for less than one year

Persistent seed bank can be short term persistent seed bank where seeds remain viable for more than one year but less than five years.

Long term persistent weed seed bank. Here seed viability extends to more than five years

Managing soil seed bank is an important aspect of weed control. The objective should be to reduce inputs into the soil and increase withdrawal from the seed bank.

Removing weeds before they set seeds is the most important aspect to be followed. Allowing the surface weeds to germinate before sowing the crop seed will help to minimise competition. In this case, care should be taken to see that surface soil is not disturbed after weed seed germination. Stale seed bed technique is adopted for removing surface seeds.

Allelopathy

Allelopthy is a natural phenomenon by which a plant species can influence the growth and development of its neighbours. The term Allelopathy was coined by the German botanist Molisch in 1937 to indicate all effects that directly and indirectly result from biochemical substances transferred from one plant to another. Hans Molish (1856-1937) also known as the father of allelopathy derived the term from the Greek words "Allelon "meaning mutual and "Pathos" meaning harm. It is confined to the mutual influence either stimulatory or inhibitory effect of one plant species on another. This is by the release of biochemical constituents into the surroundings and these chemicals can have either an inhibitory effect or a stimulatory influence on the neighbouring plant species. This phenomenon helps plants to reduce competition from neighbours and retain their space for root and shoot growth. Allelopthy has been reported in trees such as mango, teak and crops like sunflower and safflower. Allelochemicals may be present in many plant parts such as foliage, flowers to roots and bark. As leaves fall and decompose they accumulate in the soil. Mulching with such plant parts also induces allelopathic responses. Knowledge of allelochemicals can be utilised for agricultural management, such as weed control, crop protection, or crop re-establishment. Negative responses by allelo chemicals can cause autotoxicity, soil sickness, or biological invasion.

Uses of Allelopathy

Agrochemicals in Crop Production

Allelochemicals are the products of secondary metabolic pathway in plants. They can be used as growth regulators, herbicides, insecticides or antimicrobial crop protection products. They can be used as substitutes to synthetic chemicals. These natural products may not have residual toxicity though their efficiency as herbicides may be limited.

Sustainable Weed Management

Straw mulching is a sustainable method of weed control adopted by farmers (Jabran et al., 2015). Using allelopathic plants for mulching and also as ground cover provides an environment friendly option (Dhima et al., 2006). When such material decompose in the soil, weed growth is suppressed in the farmlands. It also reduces the incidence of pests and diseases. Straw mulch also helps to improve soil fertility. It is seen that the incidence of *Ipomoea* weed in corn and soybean fields can be reduced by mulching the field with wheat (*Triticum aestivum*) straw. Similarly it is seen that Rye (Secale cereale) mulch significantly reduced the germination and growth of several problematic agronomic grass and broadleaf weeds (Schulz et al., 2013)

Prevent Loss of Nitrogen from Soil

Plant roots and also residues emit Nitrogen Inhibiting Substances (NIS) into the soil. These substances reduce the emission of gaseous nitrogen from soil as oxide and improve the utilisation rate of nitrogen fertilisers. This also reduces soil and water pollution.

Encourages Crop Rotation

Continuous monoculture in a land reduces fertility and makes the soil sick. This is also due to release of allelochemicals into soil. So to improve land utilization and increase annual output of the soil, crop rotation and intercropping can be followed. Odeyemi et al. (2016) observed that the weed (*Chromolaena odorata*) can reduce nematode population in the field. In cotton, intercropping with sorghum (*Sorghum bicolor*), sesame (*Sesamum indicum*) and soybean (*Glycine max*) improved productivity of cotton and reduced incidence of purple nut grass (*Cyperus rotundus*), Iqbal et al. (2007)

References

Baker, H. 1974 . The evolution of weeds. Annu Rev Ecol Syst 5:1–24

Borland, A. M., Griffiths, H., Maxwell, C., Broadmeadow, M. S. J., Griffiths, N. M. and Barnes J. D.1992. On the ecophysiology of the Clusiaceae in Trinidad: Expression of CAM in Clusia minor L. during the transition from wet to dry season and characterization of three endemic species. New Phytol. 122:349–357.

Chauhan, B.S., Gill, G.and Preston, C.. 2006. Tillage system effects on weed ecology, herbicide activity and persistence: Areview. Aust. J. Exp. Agric. 46: 1557-1570.

Chauhan, B.S., Singh, R.G. and Mahajan, G. 2012. Ecology and management ofweeds under conservation agriculture: A review. Crop Prot. 38: 57-65.

Cousens, R. and Mortimer, M.1995.Dynamics of Weed Population.Cambridge University Press. New York.

Das, T.K., Tuti,, M..D. Sharma, R., Paul, T. and. Mirjha, P.R. 2012. Weed management research in India: An overview. Indian J. Agron. 57 (3rd IAC: Spl Issue): 148-156

de Mattos, E. A. and Lu'ttge, U.. 2001. Chlorophyll fluorescence and organic acid oscillations during the transition from CAM to C3-photosynthesis in *Clusia minor* L. (Clusiaceae). Ann.Bot.(Lond.)88:457–463.

Dhima, K., Vasilakoglou, I., Eleftherohorinos, I., Lithourgidis, A., 2006. Allelopathic potential of winter cereals and their cover crop mulch effect on grass weed suppression and corn development. Crop Sci. 46, 345e352.

Gealy, D. R., Squier, S. A., and Ogg, A. G. 1991. Photosynthetic productivity of mayweed chamomile (*Anthemis cotula*). Weed Sci. 39, 18–26. doi: 10.1017/S0043174500057805

Girija, T., Francis, R..M.,, Kumari, L., Abraham, C.T. and Zachariah,,G. 2002. Weed su ppression in upland rice: influence of physiological and morphological attributes. In Proceedings, National Symposium on Priorities and Strategies for Rice Research in High Rainfall tropics: 10-11 October, Regional Agricultural Research Station, Pattambi, Kerala Agricultural University.

Grime, J.P. 19 77. Evidence for the existe nce of th ree primary str ategies in pl ants and its relevance to ecological and evolutionary theory. Am. Nat. 111(982): 1169–1194.

Haag-Kerwer, A., Franco, A.C. and Lu"ttge.U. 1992. The effect of temperature and light on gas exchange and acid accumulation in the C3–CAM plant *Clusia minor* L. J. Expt. Bot. 43:345–352

Harrington, K.C., Thatcher, A. A and Kemp, P.D.2006.Mineral composition and nutritive value of some common pasture weeds. New Zealand Plant Protection 59:261-265

Heap I. 2014 .Global perspective of herbicide-resistant weeds. Pest Manag Sci. Sep;70(9):1306-15. doi: 10.1002/ps.3696. Epub 2014 Jan 15. PMID: 24302673.

Holm L.G., Plucknet G.L., Pancho J.V., and Herberger J. B.1977. The World's Worst Weeds. Distribution and Biology. University press of Hawaii, Honolulu.

Iqbal, J., Cheema, Z., & An, M. 2007. Intercropping of field crops in cotton for the management of purple nutsedge (*Cyperus rotundus* L.). Plant and Soil, 300, 163–171.

Jabran, K., Mahajan, G., Sardana, V., Chauhan, B. S., 2015. Allelopathy for weed con-trol in agricultural systems. Crop protection (72) 57-65.

Jayasree, P. K. 2005. Biology Management of Mimosa invisa Mart.in Kerala. Ph.D. Thesis. Kerala Agricultural .University, Thrissur

Kingman, G.C. and Ashton, F. M. 1982.Weed Science: Principles and Practices. John Wiley and Sons, New York.

Lu ̈ttge, U. 2007. Clusia: a woody neotropical genus of remarkable plasticity and diversity. Heidelberg: Springer-Verlag

Malik, R. K., and Kumar, V., 2014. Zero tillage and management of herbicide resistance in wheat. In Souvenier. Directorate of Weed Research, Jabalpur, India. pp. 64-70

Malik, R. K and Singh, S.1993. Evolving strategies for herbicide use in wheat. Resistance and integrated weed management. In Proceedings, International Symposium, Indian Society of Wee d Scien ce.18-20 Novem ber. Hisar, India, 1, 225-23 8.

Malik, R.K and Singh, S.1995.Little seed canary grass (*Phalaris minor* Retz.) Resistance to isoproturon in In dia. Weed Tec hnol ogy 9:419-4 25

Martin, A. R.and Burn side, O. C.1980. Common milk weed-weed on the in-crease.Weeds Today 11(1): 19-20

Narwal, S.S. 1994. Allelopathy in Crop Production.Scientific Publishers, Jodhpur., India. 288pp.

Narwal S.S and Jain.S.K.1994. Hans Molisch (1856-1937): The father of allelopathy. Allelopathy. J.1(1). 1-5

Nawaz, M. and S. George. 2004. Eupatorium {*Chromolaena odorata* (L.) King and Robinson. Biomass as a source of organic manure in okra cultivation. J. Trop.Agri. 42: 33-34.

Odeyemi, I. S. Afolami, S.O. and Azeez, J. O. 2016. Influence of soil properties on plant nematode population density under *Chromolena odorata* Fallow. J. Agric. Sci. & Env. 16 (1): 105-115.

Robert, H. G. and Steven, J. S. 2009.Weed Seed bank: Biology and management. Prairie Soils and crops Journal 2:46-52

Ross, M.A., and Lembi, C. A. 2008. Applied weed science: Including the ecology and management of invasive plants, 576 p. 3"Edition. Pearson Education

Salisbury, F. B. and Rose, C. W.1992. Plant Physiology. Fourth Edition. Belmont.C.A: Wadsworth. Inc.

Schulz, M., Marocco, A., Tabaglio, V., Macias, F.A., Molinillo, J.M., 2013. Benzox-azinoids in rye allelopathy-from discovery to application in sustainable weed control and organic farming. J. Chem. Ecol. 39, 154e174.

Shetty, S.V. R., Kr antz, B.A., 1 980. Weed res earch at ICRISAT (inter nat ional cro ps re- sea rch institute for the semi-arid tropics). Weed Sci. 451–454

Shiji, C.P. 2001. Relationship between weed density and yield loss in semi dry rice. M. Sc. Thesis, Kerala Agricultural University, Thrissur.

Sreepriya, S. and Girija, T.2018. Seed priming for improving the weed competitiveness in sesame. Journal of Crop and Weed, 14(2): 40-45

Stevens, O.A. 1932. The number and weight of seeds produced by weeds. American journal of Botany 19: 784-794.

Stevens, O.A. 1957.Weight of seeds and number per plant. Weeds 5: 46-55.

Suada, A. P.2015.Growth and physiology of Isachne miliacea Roth. in different soil types and its sensitivity to common herbicides. M. Sc. Thesis, Kerala Agricultural University, Thrissur.

Tillman, R. W., Scotter, D. R., Wallis, M. G., and Clothier, B. E. 1989. Water repel-lency and its measurement by using intrinsic sorptivity. Australian Journal of Soil Re-search 27:637–644.

Vavilov, Nikolai .1928. Geographische Zentren unserer Kulturpflanzen. In: Verhand-lungen des Internationalen Kongresses für Vererbung swissen schaft Berlin 1927, Supplementband Zeitschrift für induktive Abstammungs- und Vererbungslehre. pp. 342–369.

Yaduraju,N. T., and Kathirasen,R.M., 2003. Invasive weeds in the tropics.In Proc. 19th Asia -Pacific Weed Sci.Soc.Conf. Manila, Philippines, Vol.1. pp. 59-68.

Yaduraju, N.T. 2012. Weed management perspectives for India in the changing agriculture scenario in the country. Pak. J. Weed Sci. Res. 18 (Spl. Issue): 703-710.

Yaduraju, N.T. and R.M. Kathiresan, 2003. Invasive weeds in the tropics. Proc. 19th Asian-Pacific Weed Sci. Soc. Conf., Manila, Philippines. Vol. I: 59-68.Zimdahl, R. L. 2007

Weed-Crop Competition: A Review. Hoboken, NJ: John Wiley and Sons

4

Weed Control

Weeds have been a problem to agriculture ever sine man started cultivation. Weed species by and large invade cropped areas and account for large economic losses to the farmer. Weeds are economically more important than insects, fungi or other pest organisms (Savary et al. 2000, 2006). Globally, weeds cause the highest potential loss to crop production (34%), with animal pests and pathogens being less important (losses of18% and 16% respectively). The loss in production due to weeds has been estimated as 5% in developed countries. It is around 10% in developing countries and 25% in least developed countries. (Oerke, 2006).

Invasive Alien Species and Their Control

Mostly weeds are native plant flora with high adaptability to an ecosystem, their growth and spread in the ecosystem is kept in check by natural enemies of the ecosystem. Geographical boundaries have provided enough isolation for ecosystems and species to evolve. Human travel around the globe has led to the introduction of many species to new environment both intentionally and otherwise. These introduced species which lack natural enemies sometimes invade the ecosystem and try to suppress the native flora. These are known as Invasive Alien Species (IAS). According to Wijesundara (2010) highrate of introduction has resulted in unprecedented invasion of species in to new ecosystems. The process of invasion has been accelerated due to the lack of natural impediments in dispersal of these species. Alien introduced species have devastated the ecosystem and caused enormous economic and ecological damages.

Some of these plants have been introduced for specific purpose like the introduction of *Mikania*. The weed is a fast growing climber which was used as camouflage to help troop movement during the II nd world war. These plants have escaped from the point of introduction and become invasive.

Introduction of species to new ecosystem has been going on from colonial days. With frequent movement of man and material between nations, chances of introduction is much higher. Awareness about the damages caused by invasive species to the ecosystem has contributed to the establishment of many checks and regulations to prevent introduction.

Table 1: List of AIS of Weed in India

Name of the weed	Scientific name	Year of introduction	Method	Ist Report
Lantana	*Lantana camara*	1809	Plant collector	Culcutta
Giant sensitive weed	*Mimosa invisa*	1964	Cover crop	Perunna Kottayam district, Kerala
Mile a minute weed	*Mikania micrantha*	1940	Camouflage	North East India II world war
Siam weed	*Chromolaena odorata*	1960	Plant collector	Mysore (Western Ghats)
Congress weed	*Parthenium hysterophorus*	1955	Trade Wheat import	Pune
Water fern	*Salvinia molesta*	1955	Research	Veli lake Kerala
Water hyacinth	*Eichhornia crassipes*	1896	Plant collector	Botanical garden, Bengal
Algarroba, ironwood	*Prosopis juliflora*	1877	British	Tracts of S.India

A number of different specialised agencies are engaged in preventing the introduction of invasive species and also for their management and control. These include the Ministry of Environment Forests and Climate Change, the National Bureau of Fish Genetic Resources, Plant Quarantine Organisation of India and various departments of the Ministry of Agriculture.

Negative Impact of Invasive Alien Species (IAS)

Loss of biodiversity: IAS have been regarded as the second most common threat associated with species that have gone completely extinct (IUCN red list).This threat is not only for plants but also with regard to the extinction of amphibians, reptiles, and mammals.

Changes in ecosystem: IAS can change the structure and composition of the ecosystem. This inturn affects the ecosystem services with a detrimental effect on economies and human well-being. Introduction of water hyacinth *Eichhornia crassipes,* a native to South America into our ecosystem has become a threat to rice cultivation in Kuttanad area of Kerala blocked inland navigation routes and increased cases of vector-borne diseases. Financial loss incurred for its control and removal has become a burden to the local government. The fast-growing floating aquatic plant forms a dense mat on the water surface, limiting oxygen and preventing sunlight reaching the water column. This affects fish farming in the area.

Loss of habitat: Invasive species can have a number of negative impacts on the areas that they invade. Perhaps the most significant of these is the widespread loss of habitat.

Extinction of native flora: Fast growing invasive species compete for resources with the native flora and weed shift is commonly seen in most places. This can lead to extinction of native flora.

Resource crunch: Most of the IAS are fast growing and compete with native flora for resources. In many plantations they smother tree crops, especially climbers like *Mikania micrantha.*

Impact on human and animal health: Invasive species like *Parthenium* is reported to be a notorious allogeneic plant with health effects being reported from almost all countries where the weed has invaded.

Economic cost: In addition to these impacts, invasive species can also have enormous economic costs. For example threat to fisheries by water hyacinth and cost for removal of climbers like mikania from plantations. Governmental agencies are spending a lot of money on the eradication of parthenium.

Managing Invasive Species

In the last decade, there have been efforts to compile list of invasive plant species in India and to study the impacts of invasive species in different parts of the country. In 2009, the Indian Council for Forestry Research and Education set up a Forest Invasive Species Cell to develop capacities for invasive species management and to create a database on invasive species.

An integrated forest protection scheme was devised to include the management of invasive species. The last tiger census conducted by the National Tiger Conservation Authority included a survey of the distribution of a subset of invasive plants in tiger landscapes across the country. And the 12th five year plan proposed a national invasive species monitoring system.

We have a number of different legislations relating to invasive species. Some of these were enacted long before invasive species were a global concern – but have since been amended to include invasive species. An indicative, though incomplete, list includes;

The Plant Quarantine (Regulation of Import into India) Order 2003.

The Destructive Insects and Pests Act, 1914 (and amendments).

Livestock Importation Act 1898 and the Livestock Importation (Amendment) Ordinance, 2001.

Environment Protection Act 1986 and Biological Diversity Act 2002.

Methods of Weed Control

Proper crop management across different cropping systems calls for selecting appropriate combination of methods to improve productivity, reduce cost of labour, fuel, fertilisers, water and pesticide, suppress undesirable vegetation by adopting eco friendly techniques. As the magnitude of loss caused by weeds is much higher than by pests and diseases, it is important to employ long term approaches in weed control to obtain sustainable results. These include a number of agricultural practices like tillage, crop rotation, mulching, cover crops, irrigation scheduling, use of competitive varieties, type and placement of nutrients and changes in time of planting.

Cultural method of weed control can be broadly divided into preventive, mechanical and agronomic methods.

Preventive Method

Preventing introduction of new weed species into a cropped ecosystem should be the first priority. This can be achieved by proper monitoring of inputs. Removal of weeds before seed set is another approach to reduce the weed seed input into the soil seed bank. Techniques to be followed to prevent introduction of weeds into an area include the following

1. Use of certified seeds for cultivation. Do not use seeds from endemic areas
2. Adopt strict certification standards in seed production
3. Do not feed viable weed seeds to livestock, adopt ensiling methods to reduce seed viability
4. Composting reduces seed viability when the temperature goes up in the compost pits. Adopt proper composting techniques
5. Clean the machinery before transporting to different locations . This is to prevent dispersal of weed seeds through machinery
6. Employ stale seed bed technique to exhaust the soil seed bank of weed seeds
7. Keep non-cropped areas like irrigation channels, bunds and fence line clean and free of weeds
8. Prevent establishment of new weeds through regular vigil
9. Do not use residues from the field as packaging material
10. Follow legal and quarantine measures.

Mowing and Slashing

This is done to prevent weed seed production and dissemination through concurrent control of the existing weeds or wild vegetation usually under non-cropped situation such as canals, bunds, farm roads, parks, pastures and lawns by using tractor, battery or electricity operated mowers or choppers. This operation should be carried out before weeds set seeds. The process is more effective on annual and erect growing weeds rather than perennials with creepy nature. Mowing is done with sickle mowers and rotary mowers. It is commonly used in plantations, hedges, pathways, channels and on bunds.

Water Submergence

Many weeds cannot germinate under flooded condition. Flooding is practiced in rice cultivation. Maintaining 5cm of water throughout the growing season may be adequate to control most of the weeds. Flooding prevents weed seed germination, root respiration of germinated seeds, kills them by reducing oxygen supply for germination and growth leading to suffocation and death. If the land preparation is faulty it may not be so effective. Weeds may grow on exposed areas. It is seen that flooding controls grasses and sedges better than broad leaved weeds. Once the weeds get established it is difficult to control by this method. In transplanted rice the most critical period is from transplanting to tillering stage which comes to around 12 to 35 days depending on the duration of the crop variety. Nature and extent of weed growth is found to vary with flooding depth. Under shallow flooding of 2 cm height there will be more sedges than broadleaf weeds and grasses. While 5cm of standing water, was found to suppress both grasses and sedges. Saturated condition favours growth of predominant weeds like barnyard grass (*Echinochloa* spp) and sedges like *Fimbristylis* spp.

Mulching

Mulch is a protective covering placed on the soil surface to prevent weed emergence. Mulching can be done with different materials like crop residues, straw, leaves, paper, plastic film or gravel. In addition to weed control it provides erosion control, conserves moisture, reduces temperature fluctuations, increases soil organic matter content and improves soil structure. Plastic mulches have become very popular in vegetable cultivation. Due to the adverse effect of plastic mulches on the environment eco friendly alternatives like coir pith mulches are also available. One of the challenges of using plant residues for mulching is the problem of termite attack on the crop.

Mechanical Methods

Tools and implements used in mechanical method of weed control are mostly operated by human labour, animal and tractors. These included

Hand Pulling

This is the most efficient method of weed control. It is important that expertise is required in the identification of weeds. It is difficult in broadcast method of sowing. Availability of cheap labour is a pre requisite. Under the present condition where cost of labour is very high this technique is not economically viable.

Manually Operated Weeding Tools

These include spades or chopping hoe, push pull type or long handled weeders, rotary weeders and hoe type weeders. Depending on the soil type different types of hoes are used in different regions these include hand hoes, weeding hook or fork hoe.

Hand Hoe

Hand hoes can be effectively used in field condition where row planting is adopted. Hoes are used to cut below the crown of the plants causing their complete destruction in case of annual weeds, but perennial weeds with runners and stolon will pose a problem for this method of weed control.

Rotary Weeder

Rotary weeders are effective when row planting is done. These equipments are effectively used in the garden lands and upland cultivation of crops like rice, maize etc.

Sensor Based Mechanical Weeders

Since 1980 a number of sophisticated sensor attached instruments are available in the market Camera-steered hoes with a hydraulic side shifting control for row crops are robust and reliable and they are now widely available from different manufacturers

Robotic Weeders

Robotic Weeder will roam around the garden systematically identifying and killing weeds. This eliminates the need for weeding which is tedious and is the most unpopular part of gardening. The robot is fitted with caterpillar tracks to cope the uneven terrain of gardens. Robotic Weeder is water proof and is also fitted with a solar panel so that it can charge 'on the go' and reduce its impact on environment. This type of weeder has three sensors to build up a 3d map of the garden to navigate it. There is a proximity sensor, a camera and an infra-red sensor. Included with the Robotic Weeder is a base unit (charger) and two types of waterproof electronic plant tags. Cultivated plants in the garden should be attached with identification tag so the robot knows they are not weeds.

Weed Warden

This is a comparatively low-cost (<$200) plant detection sensor that can be mounted on rovers or tractors. The Weed Warden uses an open source multispectral sensor to detect live vegetation and sends a logic signal that could trigger a weed removal system such as a sprayer or mechanical tillage when vegetation is detected.

Drones

Drones are being successfully used for spaying herbicides in plantations, forest area with difficult terrain and also on water bodies to control floating weeds.

Physical Methods

Land Levelling

Levelling ensures uniform distribution of farm resources such as seed, water, fertilisers and herbicides. It is estimated that proper levelling can reduce weed pressure and infestation by 10 to 20%.This ensures uniform growth of the crop and also weeds. When the land is uneven it accounts for staggered germination of weed seeds. Control measures become more effective when the weed growth is uniform.

Flaming

Flaming the field after harvest of the crop is a common practice in many regions. This helps to eliminate weed seeds as well as remanence of pest and disease from the field. It is also employed to eliminate vegetation from non cropped areas. In foreign countries flame throwers are used for selective control of green and newly emerged weeds in rows like cotton, onion, vegetables etc. However this is not a recommended practice as it contributes to air pollution.

Dredging

In aquatic system to remove aquatic weed along with their roots and rhizomes dredging is done with heavy chains and nets attached to animals or boats. Now earth movers are also used for dredging in shallow waters.

Agronomic Practices

A number of agronomic practices are used to reduce weed infestation in crop fields. These include

Tillage

Tillage is the most widely accepted cultural operation adopted worldwide to attain a good seed bed for seed germination and better root growth. It helps to reduce weed population by exhausting soil seed bank. It reduces population of perennial weeds by exposing the vegetative propagules so as to reduce their viability. Tillage operations are conducted to prepare a good seed bed. This brings the weed seeds to the surface and induces germination. During the intercultural operation the seedlings are killed. In case of perennial weeds, tillage operations injures the root and stem and weakens their regenerative and competitive capacity.

Cover Cropping

Cover crops can be used in fallow lands or in plantations. Cover crops grow very fast and create unfavourable environment for weed growth there by preventing their establishment and seed set. This can be by physical suppression, resource consumption or through allelopathy. They also improve soil health and productivity of the succeeding crop. They are mainly used in rubber plantations for weed suppression.

Crop Rotation

According to Malik et al 2002 weed communities in a continuous cropping system tend to have greater total densities. A solution for this is to diversify the cropping system. Crop rotation prevents the dominance of any single weed and also delays the evolution of herbicide resistant weed species in the ecosystem. Control of *P. minor* in wheat was achieved by replacing wheat with other crops like berseem, rye, winter maize sugar cane etc. during rabi season. These crops help to smother the weed as the growth habits are different. Wild oats (*Avena fatua*) was completely eliminated from wheat by growing berseen 3-4 times (Gill et al 1985). Problem with *P.minor* was overcome by shifting to the cultivation of sunflower in wheat fields which provided better incentive to the farmers (Malik et al.2002). Crop rotation breaks the cycle of crop associated weeds that dominate in monoculture. Crop rotation changes the composition and abundance of weed species and employs varying patterns of resource competition. Seeds of problem weeds that are accumulated in the soil seed bank gets exhausted by crop rotation.

Competitive Cultivars

According to Zhao (2006) competitiveness of a variety and weed tolerance are two important components of cultivar weed competition. Most of the present day high yielding varieties have less tolerance to weeds. These varieties are more sensitive to fertiliser input and respond well to better nutrition and management practices. It is important to identify varieties with better competitive characters. These include faster early growth and establishment. When plant growth is faster during the initial stages after germination they cover ground quickly preventing weed germination. So pressure from weeds will be much lower. Hence it is important to incorporate characters like early emergence, rapid creation of a dense canopy, increased plant height, early root growth and increased root size in evolving weed competitive varieties of annual crops.

Screening of twenty six upland rice varieties for weed competitiveness revealed that rice varieties can be classified into three groups. Varieties that can completely smother weeds but the productivity of these varieties is generally low as dry matter partitioning is more towards vegetative growth. Varieties that showed moderate tolerance to weeds these are the potential varieties that can be improved for productivity characters. The third group contained varieties that are completely smothered by weeds. Assessment of physiological and morphological traits of these varieties revealed that early vigour, plant height and plant architecture are the major factors that decided competitiveness of varieties. (Girija et al. 2002)

Allelopathic traits have been reported in rice varieties by Fuji in 1992. Incorporating such traits into high yielding varieties can also improve competitiveness of rice varieties.

Competitive ability of wheat crop was found to be associated with height of the plant and early canopy cover. As these plants intercept more light, it reduces weed germination and growth (Balyan and Malik,1991)

Method of Sowing

Adopting closer row spacing of field crops gives an upper hand for the crop over the weeds. Thick crop stand smothers growth of weeds and also prevents germination of weed seeds. Bidirectional sowing reduces weed growth due to better distribution of crop stand and development of canopy structure which covers weeds effectively.

Seed Rate

Increasing seed rate for a thicker stand or increasing plant population in unit area gives a competitive advantage over weeds. This is adopted by farmers in direct sown rice as a technique to suppress weeds. The per plant productivity is found to be lower, but productivity from unit area is found to improve. Moreover effective weed control is also possible. In wheat it was reported that increased seed rate supplemented with herbicide application lead to a marked reduction in the dry matter of *P minor* (Malik et al., 1984)

Fertiliser Application

Placement of fertilisers closer to the root zone of the crop will help the crop to gain higher growth over weeds. In broad cast application of fertilisers, it is important to weed the field before fertiliser is applied. Basal dose of nitrogen helps early establishment of the crop. This gives the crop a competitive advantage over weeds.

Stale Seed Bed

This is a very effective technique which can be followed in field crops like rice and wheat. The field is irrigated either by early rains or through irrigation facility one or two months before sowing. This induces weed germination. The germinated flushes are then destroyed by herbicide or by cultivation practices. This method is most effective in cases where the germination of the crop and weed seed is synchronised. However this technique will not lead to complete elimination of the weed but only reduce the population. In very severe cases of weed infestation in the field (e.g. wild rice) the processes is repeated twice in a season before sowing to get a better control.

Soil Solarisation

In high value crops like cardamom vegetables and in nurseries, soil solarisation is a good practice to eliminate/reduce weed seeds from the soil seed bank. After land preparation, the field is covered with 25-50μ transparent polyethylene sheet during the hottest period with moist soil to trap the solar radiation. Moisture in the soil helps to conduct the heat and sensitise seeds. Normally seeds are heat resistant when dry, higher soil temperature under moist condition proves lethal to weed seeds. As the cost is much higher in this technique, it is recommended before planting in high value crops. The process also effectively controls pests, nematodes and weeds. Transparent polythene is used to transmit most of the solar radiation and heat to the soil. To conduct heat to lower layers it is recommended to continue the process for four weeks or more to get control at desired depths.

Response to solarisation was found to vary with weed species. Most of the annual species of tropical region were found to be sensitive while many perennial weeds are resistant.

Conservation Agriculture

The concept of conservation agriculture has been developed to reduce land degradation and ensure sustainable crop production. Change from conventional to conservation farming practices which completely eliminate tillage practices (Zero tillage) has accounted for a weed shift in many farm ecosystems. Reduction in tillage intensity and frequency and adoption of zero tillage has generally increased weed infestation. However some researchers suggested that though the initial weed density was higher in zero tillage system in subsequent years the number dropped drastically due to increased seed predation, decomposition and germination under detrimental environmental conditions (Buhler 2002, Malik et al., 2015). Considering the diversity of weed problems in conservation agriculture, no single method alone can be employed for effective control of weeds. So a combination of different strategies was found to be more effective. Minimum tillage has found to significantly reduce infestation of *P minor* in wheat field in North West India(Malik,2002).

Integrated Weed Management

The term IWM was first introduced by Shaw in 1982.Integrated weed management (IWM) is regarded as a weed management system that utilises all suitable techniques and methods in a compatible manner to maintain weed population at levels below those causing economic injury. This technique of using many little hammers to meet a problem, prevents weeds from getting adapted to a situation and leads to long term control and sustain diversity in the ecosystem. Liebman et al. (2001) brought in the concept of ecological weed management. Instead of weed control, the new concept is to focus on weed management. According to Harker and O'Donovan (2013). Integrated weed management (IWM) can be defined as a holistic approach to weed management that integrates different methods of weed control to provide the crop with an advantage over weeds.

Under the present scenario where labour shortage and high wages have become major challenges in crop production, herbicides have received a lot of popularity with farmers, due to their simplicity in application, affordability and effectiveness. However, relying too much on a few herbicides has led to an increase of weed species that are not effectively controlled with herbicides resulting in the emergence of herbicide-resistant biotypes. Integrated Weed Management (IWM) has been accepted as an effective strategy for sustainable and long term control of weeds. It combines various methods to reduce or eliminate the effect of weeds on crop production over time, using a combination of practices that are most effective for solving specific weed issues. It actually combines all available methods that will best solve the problem. In many cases with conventional crops, herbicides are also part of this. It is not a replacement for herbicides.

To implement IWM it is important for the farmer to acquire knowledge on aspects such as type of weeds growing in a cropping system and its ecological attributes like time of emergence, duration of the weed, critical period of crop weed competition, cultural operations to be adopted in the field, spacing of crop, chemicals suitable for control of weeds in the crop, crops suitable for intercropping and for rotation. Mulches suitable for the crop bio control agents that can be included

Adopting more than one technique for weed management has the potential to restrict weed populations to manageable levels, reduce the environmental impact of individual weed management practices, increase cropping system sustainability, and reduce selection pressure for development of weeds resistant to herbicides (Buhler, 1999; Swanton and Weise 1991). This technique is like driving many little nails to a plank so as to improve the stability of the plank.

A judicious choice of available techniques is warranted for effective results. This is bound to vary with the crop, season of cultivation, method of cultivation, available options of control on the targeted weed species. All the currently available techniques for weed management such as prevention, depleting the soil seed bank, cultural control, mechanical control and chemical control can be applied as per requirement for achieving the desired level of control in the field.

Weedy Rice Control – A Case Study

In rice cultivation, control of weedy rice is a problem due to the close similarity between rice plant and the weed as they belong to the same genera. None of the herbicides available in the market can be used for the selective control of the weed. A three pronged technology was suggested for its control (Jose, 2015) which emphasizes on weed seed depletion from the soil seed bank, controlling the germination of the weed from the existing seed bank and thirdly preventing the replenishment to the seed bank.

The ear head and grain characters of the weed showed high variability. Presence of staggered dormancy is the biggest challenge. The initial germination (just after harvest) ranged from 8-12%. Above 50% of the seeds germinated after 5 months of storage, 5-10% seeds germinated during the intervening period.

Chaff percentage was very high (up to 70%) in all the variants collected. These factors made the elimination of weedy rice from soil seed bank a laborious task which may take a number of years.

For depleting the weed seeds from the existing seed bank, stale seed bed technique was employed. In dry sown situation, after land preparation weed seeds were allowed to germinate and after two weeks, to destroy these weeds a broad spectrum herbicide like paraquat or glyphosate was sprayed. Rice seeds were then sown without much disturbance to the soil. In wet seeded rice after seed bed preparation, field is drained for 2-3 weeks to enable weeds to germinate. Glyphosate is then sprayed and after two weeks the field is re-flooded for ten days till the weeds decay. The field is than drained and germinated rice seeds were broadcasted. A single stale contributed to 50 % reduction in weed seeds while 2 continuous stales reduced the seed bank by 75-80%.

Second method was to control the germination of the weeds from the existing seed bank with a pre emergence spray of oxyfluorfen @ 0.3- 0.4 g ai/ ha. This is sprayed with 2cm water in the field. The water is then allowed to evaporate so that the herbicide forms a thin film on the soil surface this controlled germination of the weed seeds from soil seed bank. Germinated rice seeds were sown in the field with minimum disturbance to the soil.

Third method is to prevent replenishment to the soil seed bank by killing the panicles of weedy rice as and when they emerge. This will also ensure clean paddy seed production. In the case of rice varieties which have duration of 110 to 125 days, panicles of weedy rice emerges about 60 days after sowing when rice cultivars will be in the vegetative stage. At this stage, there will be a marked difference between the height of the crop and the weed so the panicle of weedy rice can either be cut and removed or dried by direct contact application. This can be achieved manually or with the help of a weed wiper. KAU weed wiper is an equipment developed for the purpose. Herbicides such as glyphosate, glufosinate ammonia or paraquat can be used @ 100ml/L.of water. The coverage and efficiency of application depends on the experience of the person operating the equipment.

Employing all these techniques simultaneously for two to three seasons helped reduce the menace of weedy rice in the field.This was successfully demonstrated by the farmers of Kuttanad.

References

Balyan, R.S. and Malik, R.K. 1991. Competitive ability of wheat cultivar with wild oat (*Avena ludoviciana*).Weed Science 39(2):154-15.

Buhler, D.D. 1999. Expanding the context of weed management. New York: Haworth Press. 289

Buhler, D.D. 2002. Challenges and Opportunities for Integrated Weed Management. Weed Science, 50, 273-280. http://dx.doi.org/10.1614/0043- 745(2002)050[0273:AIAAOF]2.0.CO;2

DWR. 2015. Vision 2050, 32p. Directorate of Weed Research, Jabalpur, Madhya Pradesh

Fuji, Y.1992. The potential biological control of paddy weeds with allelopathic effect of some rice varieties. pp. 305-320.In: Biological Control and Integrated Management of Paddy and Aquatic Weeds in Asia. National Agricultural Research Centre, Tsukuba, Japan.

Gill,H.S, Brar, L.S. and Paul, S.1985. Effect of different crops on the growth and development of wild oats. In: Proceedings of the Annual Conference of Indian Society of Weed Science, Gujarat Agricultural University, Anand, 20-230February.

Girija, T., Francis, R.M., Leena, S., Abraham, C.T., Zachariah, G. 2002. Weed supression in upland rice influence of physiological and morphological attributes.In: Proceedings of the National Symposium on Priorities and Strategies for Rice Research in High Rainfall Tropics:10-11 October Regional Agricultural Research Station, Pattambi, Kerala. India

Harker and O'Donovan.2013. Weed control, weed management, and IWM .Weed Technology 27:1–11

IUCN and MENR 2007. The 2007 Red list of threatened fauna and flora of Sri Lanka.The World Coservation Union, and the Ministry of Environment and natural Resources,Colombo,Sri Lanka.148pp

Jose, N, Abraham ,C.T., Mathew, R. and Leenakumari, S. 2015a. Management of weedy rice by novel wick applicator. Proceedings of 25th Asian-Pacific Weed Science Society Conference, 13-15 October, 2015, Hyderabad,India.

Jose. N. 2015. Biology and Management of Weedy Rice. Ph.D. Thesis, Kerala Agricultural University, Thrissur. 230 p.

Liebman, M., Mohler, C.L. and Staver, C.P. 2001. Ecological management of agricultural weeds.Cambridge University Press, Cambridge, UK.

Malik, R.K., Balyan, R.S. ans Bhan, V.M.1984. Integrated control of Phalaris minor in wheat. Indian J. Of Weed Science 16(1):52-55.

Malik, R.K., Kumar, V., Yadav, A. and Mc Donald, A.2015. Weed science and sustainable intensification of cropping systemin South Asia. In: Proceedings Vol I(Plenary & lead papers). 25th Asia Pacific Weed Science Conference on Weed Science for Sustainable Agriculture, Environment and Biodiversity(Eds. AN Rao and NT Yaduraju),13-16 October,2015,Hyderabad India: Vol I:67-78

Malik, R.K., Yadav. A., Singh, S., Malik, R.S., Balyan, R.S., Banga, R.S., Sardana, P.K., J[ai]pal, S., Hobbs, P.R., Gill, G., Singh, S., Gupta, R.K. and Bellinder, R. 2002. Herbicide Resistance Management and Evolution of Zero Tillage-A Success Story, CCSHAU Research Bulletin, 43p.

Oerke, E. C. (2006). Crop losses to pests. Journal of Agricultural Science, 144,31–43.

Savary, S., Teng, P. S., Willocquet, L., & Nutter, F. W., Jr. 2006. Quantification and modeling of crop losses: a review of purposes.Annual Review of Phytopathology, 44,89–112

Savary, S, Willocquet, L., Elazegui, F. A., Castilla, N. P., & Teng, P. S. 2000. Rice pest constraints in tropical Asia: quantification of yield losses due to rice pests in a range of production situations. Plant Disease, 84, 357–369.

Shaw, W.C. 1982. Integrated Weed Management Systems Technology for Pest Management. Weed Science, 30, 2–12. http://www.jstor.org/stable/4043545

Swanton, C.J., Weise, S.F. 1991. Integrated weed management: the rationale and approach. WeedTechnol 5:648–656

Wijesundara, S. 2010.Invasive alien plants in Sri Lanka-Strengthening capacity to control their introduction and spread ,pp.27-38 In: B. Marambe .P. Silva, S. Wijesundara and N. Atapattu (eds.). Biodiversity Secretariate of Ministry of Environment, Sri Lank.

Zhao, D. 2006. Weed competitiveness and yield ability of aerobic rice genotypes. Ph.D. thesis. Wageningen University, Netherlands.

5

Biological Control

Deliberate use of natural enemies to control invasive pest in an ecosystem is known as biocontrol. To control invasive weed species, selective pathogens and weed seed predators are employed to reduce the population of the weed to non competitive numbers rather than total eradication of the weed species (Appleby, 2005) These can be insects, herbivorous fish, other animals, disease causing organisms or even competitive plants. This method is comparatively safer to the ecosystem, is cheaper and has long lasting impact. Difficulty in finding host specific bio-agent with minimum adverse effect on other crop and the environment is a challenge. A number of bioagents have been tested in different parts of the world and there are a few success stories.

The cactus *Opuntia vulgaris,* had become wide spread in India in the absence of its South American natural enemies. In 1795 cochineal insect, *Dactylopius ceylonicus* (Green) was introduced to northern India from Brazil as a source of carmine dye. It was observed that this introduced insect could control *Opuntia vulgaris.* Once the value of the insect as a biocontrol agent was recognised in 1836-1838 it was introduced to south India to control the weed. This was the first successful, intentional use of an insect to control a noxious plant. Later on in 1865, *D. ceylonicus* was transferred from India to Sri Lanka which resulted in the successful control of *O. vulgaris* throughout the island (Goeden 1978, Moran & Zimmerman 1984).

Most successful control of noxious weeds by biological control was reported from Australia. It was in the control of prickly pear cacti, *Opuntia inermis* de Candolle and *O. stricta* Haworth. After screening 150 species of insects which were found to feed on the cacti *Cactoblastis cactorum* was found to be most successful. It was then introduced to Australia in 1926. Almost 90% of the original stands of *O inermis* and *O stricta* were destroyed by 1934 through larval feeding by this moth, supplemented by airborne, soft-rot bacteria for which the borers provided entrance wounds into infested plants. Virtually complete control of the cacti was achieved in Queensland and northern New South Wales nearly 24 million ha of formerly infested land was restored for agricultural use (Dodd 1940, Goeden 1978, Moran & Zimmerman 1984).

In 1972, in India an attempt was made to control the aquatic weed *Salvinia molesta* by biological agents. The grass hopper *Paulinia acuminata* was first tried but this was not very successful. Then again in 1983 sands weevil *Cyrtobagous salviniae* was released in different parts of Kerala. This insect successfully established in all the released regions. Within four to six months yellowing of the weed mat was observed and within 10 to 12 months time nearly 99% of the weed was cleared from the aquatic regions of the state (Joy et al., 1985).

Characteristics of An Ideal Biocontrol Agent

Any living organism used to stop the invasive nature of a weed is a biocontrol agent. As invasive species are non native plants, introduction of their natural enemies from the native ecosystem is the adopted procedure to control the weed species. When choosing a biocontrol agent the host specificity of the species should be ascertained. Feeding tests are conducted in the laboratories to ensure the specificity of the target weed species and also that other crop plants of the ecosystem will not serve as feed for the introduced species. High fecundity is another desirable character which ensures rapid multiplication and spread of the biocontrol agent. Ability to locate the weed species and gregarious nature of the agent are also desirable attributes. The organism should be able to sustain in the ecosystem and bring about an ecological balance.

Table 1: List of Bio control agents used in weed control

Weed species	Scientific name	Biocontrol agent (Insects)	Class
Water hyacinth	*Eichhornia crassipes*	*Orthogalumna terebrantis* Wallwork *Neochetina burchi* Hustache *N. eichhorniae* Warner	Mite Weevil Weevil
Alligator weed	*Alternanthera philoxeroides*	*Cassida syrtica* Boheman Aqasicles *hygrophila*	Beetle Flea beetle
Cocklebur	*Xanthium strumarium*	*Mecas saturnina* Lec. *Nupserha antennata* Gah.	Seedfly Beetle
Water lettuce	*Pistia stratioses*	*Namanqana pectinicoris* Hampson	Moth
Lantana	*Lantana camara*	*Epinotia lantana* Busck. *Octotoma scabripennis* Guerin-Meneville	Moth Leaf miner Beetle
Willow primrose	*Ludwigia adcendens*	*Tyloderma affineWibmer* *Liothrips ludwigi Zamar*	Weevil Thrips
Siam weed	*Chromolaena odorata*	*Pareuchaetes pseudoinsulata* Rego Barros	Moth

Weed species	Scientific name	Biocontrol agent (Insects)	Class
Carrot weed	*Parthenium hysterophorus*	*Zygogramma bicolorata* Pallister	Mexican beetle
Nut sledge	*Cyperus rotundus*	*Athesapeuta cyperi* Marshall	Nut grass weevil
Hydrilla	*Hydrilla verticillata*	*Hydrellia pakistanae* Deonier	Leaf mining fly

Bio-control is mainly adopted for control of Invasive alien species which have become invasive due to the absence of natural enemies. In such cases introduction of natural enemies from their native ecosystem creates a natural, self-sustaining balance between the weed and the natural enemy in the new ecosystem. Once the population of the bio-control agents gets established in the new ecosystem, control becomes self-perpetuating and self-regulating as the agent becomes part of the region's ecology. Monitoring an agent's population dynamics and impact is an important part of a bio-control strategy. In cases where the system is successful, the bio-control agent will reduce weed abundance, density and impact. To reach a sustainable balance and for success to be visible it may take five to ten years. Result may vary from area to area. Bio-control has been observed to be successful in grazing areas, inaccessible locations like rocky areas, deserts and wood lands, in aquatic situation which are inaccessible and areas where chemical control may be too expensive or not permitted.

Advantages and Limitations of Bio-Control

Bio-control is a green alternative to chemical or mechanical means of weed control. This method is specific to target weed. The system is self-sustaining as the natural enemies sustain themselves after introduction and so the control is almost permanent. It brings an ecological balance in the field and curtails the invasive nature of the weed. The system is cost effective in the long run as no further introduction is required.

Bio-control agent is specific, it controls only one species of weed. Some introduced species can disrupt the food chain if they start feeding on other crop plants in the ecosystem. The process of weed control is very slow as the bio agent takes time to bring the weed to a manageable level. Moreover, complete eradication of the weed is not possible as the pest requires the host for survival. The upfront cost of planning and developing a successful biocontrol is very high especially during the developing stages.

Bioherbicides

Bioherbicides are otherwise known as myco herbicides, as most of the bioherbicides contain spores of fungi. They are artificially cultured pathogens used as sprayable formulations like chemical herbicides. These can be both specific or non-specific. They may be selected from native places of the weed or even from other areas. The major problem with bio herbicides is that they are active only on the current population. There is no perpetuation of biogent on the weed or the location. So repeated sprays are required to control weeds. Bio agents are costly and there is no guarantee of total weed control.

Table 2: Commercial bioherbicides and target weeds

Product	Content	Target weed
De Vine	Liquid concentrate of chlamydospores of *Phytophthora palmivora*. Affects roots of weeds (liquid suspension) Ist bio-herbicide registered in USA	Strangler vine or Milk weed vine (*Morrenia ordorata)*
Collego	*Coleotrichum gleosproriodes* anthracnose causing fungal pathogen (Wettable powder)	Ameican joint vetch (*Aeschynomene virginica*) in rice and Soybean
Bipolaris	Fungal sporesof *Bipolaris sorghicola*	Johnsons grass (*Sorghum halepense*)
Lubao 2	*Colletotrichum gleosporoides* spp. *cuscuta*	Cuscuta
Phoma macrostoma Montagne 94-44B	Microcidins from Fungal pathogen *Phoma macrostoma*	Canada thistle and Dandelion and also to control a broad spectrum of broadleaf weeds in established turfgrass
Organo -So;	AI is composed of five spp of lactobacilli, citric acid and lactic acid	Partial supression of white and red clovers (*Trifolium.repens* and *T.pretense*), wood sorrel(O*xalis* spp.) in lawns
Opportune	*Streptomyces acidiscabies* causal organism of scab disease of potato contains killed cell of the organism and phytotoxin thaxtomin A	Broad range of annual grasses, broad leaf weeds and sedges in orchards, legumes and vegetables.
SolviNixLC	Ist herbicide containing a plant virus as active ingredient TMGMV U2 labelled in USA	Soda apple (*Solanum viarum*) and other Solanaceous weeds
Bialophos	Fermentation product of *Streptomyes hygroscopicus* (Toxin)	General vegetation non-selective herbicide

Limitations of Bioherbicides

Utility of bioherbicides is limited as each herbicide is specific to a particular weed species. Controlling one species alone in a cropped field with multiple weeds will not solve the problem. It is essential to develop a consortium of bio herbicides to control all major weeds in a crop. Moreover these should be compatible with other commonly used herbicides so that they can be applied as a cocktail.

References

Appleby, A. P. 2005. A history of weed control in the United States and Canada-a sequel. Weed Science 53: 762-768.

Ayyar, T. V. R.1931. The Coccidae of the prickly-pear in south India and their economic importance. Agric. & Livestock in India 1: 229-37.

Bailey, K. L., Pitt, W. M., Falk, S., Derby, J. 2011.The effects of Phoma macrostoma on nontarget plant and target weed species. Biological Control ;58:379–386.

Charudattan, R. And Hiebert, E. 2001. A plant virus as bioherbicide for tropical soda apple, Solanum viarum. Outlooks Pest Manage, 18:167-171

Crawley, M. J. 1989.The successes and failures of weed biological control using insects. Biocontrol News Inf.10:213-223

DeBach, P.1974. Biological Control by Natural Enemies. Cambridge University Press, London & New York. 323 p.

Dodd, A. P.1960. Biological control investigation projects in Queensland. 3rd Austral. Weed Conf. Proc. 1, Paper 4. 27 p.

Duke, S.O. and Dayan, F. E. 2011. Modes of action of microbially produced phytotoxins.Toxins. 3: 1038-1064

Goeden, R. D. & Andrés, L. A. 1999. Biological control of weeds in terrestrial and aquatic environments. In: Bellows, T. S. & T. W. Fisher (eds.), Handbook of Biological Control: Principles and Applications. Academic Press, San Diego, New York. 1046 p.

Goeden, R. D., C. A. Fleschner & D. W. Ricker. 1967. Biological control of prickly_pear cacti on Santa Cruz Island, California. Hilgardia 38: 579-606.

Goeden, R. D., C. A. Fleschner & D. W. Ricker.1968. Insects control prickly pear cactus. Calif. Agric. 22: 8-10.

Goeden, R. D. 1978. Part II: Biological control of weeds, p. 357-545. In: C. P. Clausen (ed.), Introduced Parasites and Predators of Arthropod Pests and Weeds: A World Review. U. S. Dept. Agric. Handb. No. 480.

Hallett SG.2005. What are bioherbicides? Weed Sci.53: 404-415

Hamlin, J. C.1924. Biological control of prickly pear in Australia, contributing efforts in North America. J. Econ. Ent. 17: 447-60.

Hubbard M., Hynes RK., and Bailey KL. 2015. Impact of microcidins produced by Phoma macrostoma, on carotenoid profiles in plants.Biol.Control.89:11-22

Joy, P.J., Sathersan, N.V.,and Lyla, K. R.1985. Biological control of weeds in Kerala. In Proceedings of Natl. Sem. Entomoph Ins., Calicut, pp 247-251

Kunhi-Kannan, K. 1928. The introduction of a new insect into Mysore. Mysore (India) Agric. J. 8: 141-45.

Kunhi-Kannan, K.1930. Control of cactus in Mysore by means of insects. Mysore (India) Agric. J. 11: 94-8. cd 4cd cde

LactoPro-Tech.2015.Organo-Sol. Liquid lacto-Fermented Herbicide.

Moran, V. C. & Zimmerman, H. G. 1984. The biological control of cactus weeds: achievements and prospects. Biocontr. News Info. 5: 297-320.

Online: http://msds.plantprod.com/document/949.

Rao, V. P., Ghani, M.A., Sankaran, T.and. Mathur, K. C. 1971. A review of the biological control of insects and other pests in south-east Asia and the Pacific Region.Commonw. Inst. Biol. Contr. Tech. Commun. No. 6: 1-149.

Smith, J., Wherley, B., Reynolds, C., White R, Senseman S, Falk S.2015. Weed control spectrum and turf tolerance to bioherbicide phoma macrostoma. International Journal of Pest management;61:91–8.

Stubbs, T. L. and Kennedy, A.C. 2012. Microbial weed control and microbial herbicides. pp.135-166 In: R. Alvarez-Fernandez (ed.). Herbicides - Environmental Impact Studies and Management Approaches. In Tech, Rijeka, Croatia.

Sushilkumar. 2006. Economic benefits in biological control of Parthenium by Mexican beetle, Zygogramma bicolorata Pallister (Coleoptera: Chrysomelidae) in India. Annals Entomol. 24 (1&2): 75-78.

Sushilkumar and Ray. P. 2011 Evaluation of augmentative release of Zygogramma bicolorata Pallister (Coleoptera: Chrysomelidae) for biological control of Parthenium hysterophorus. L. Crop Prot.30: 587-591

Wolfe, J. C., Neal, J. C., Harlow, C.D. 2016. Selective broadleaf weedcontrol in turfgrass with the bioherbicides Phoma macostoma and thaxtomin A. Weed Technology. 30: 688–700

6

Herbicides: Advantages and Limitations of Herbicide Usage in India

Herbicides are chemical substances that are capable of either killing or injuring plants and so used for elimination of unwanted plants. In Latin, the term 'herba' means plant and 'caedere' means to kill. There are many reasons for the increased use of chemicals for weed control. Shortage and high cost of agricultural labour prevent economic control of weeds by manual and mechanical means. Benefits offered by herbicides include efficient weed control, higher crop yield, better soil, water and energy conservation and higher income.

History of Herbicides

Time and energy spend by man to remove weeds from cropped areas prompted him to look for cheaper alternatives. In 1896, French grower found that *Sinapis arvensis,* a weed in grapes, was also controlled when $CuSO_4$ was sprayed to control diseases on vines. French growers also found that H_2SO_4 gave selective control of weeds without injuring the crop. The first organic herbicide DNOC was patented by George Truffaut and K. Pastac in 1932 for use as selective herbicide in cereals. Discovery and development of synthetic auxin herbicide, 2,4-D in early 1940 revolutionized agriculture in North America and Europe. Systemic action of auxins helped to kill underground parts of plants.

The potential of such chemicals led to the development of numerous organic herbicides. After Second World War, Jeolott Hill Resarch Station and Rothamstad Experimental station used phenoxy acetic acid like MCPA, 2,4-D and 2,4.5-T for weed control and this started a new era in weed control. In 1955, S-trazine compounds were reported. Substituted Ureas which inhibited photosynthesis was discovered in 1950s. Designing herbicide molecules specifically tailored to inhibit specific enzyme reaction followed with Monsanto's glyphosate in1974. Thereafter, around 2000 different herbicide molecules of 15 different modes of action have been introduced in the global market. Most of the new generation chemicals in the market have unique chemistry and are at times specific to weed species. There is a need to study the toxicology and environmental impact of these chemicals.

Herbicide Usage in India

There are more than 500 pesticide formulations registered for use in the country as on 01-10-2022 under section 9 (3) Insecticides Act, 1968. India is the second largest producer of pesticides in the world and has emerged as the 5th largest exporter of pesticides after China, USA, Germany and France. However, India accounts only for 1 % of the global pesticide consumption. Though the per hectare consumption is relatively low, indiscriminate use (in terms of quantity and timing) of pesticides in the country is a serious issue. Synthetic insecticides form the largest portion of pesticide consumption in India (60%), followed by fungicides (18%) and herbicides (16%). Rice and wheat are the major crops where herbicides are mainly used for weed control (Sharma et al.,2018).

Advantages of Herbicide Use

Pre- plant and pre- emergence herbicides help the crop to grow in a weed free environment with minimum competition. Herbicides improve crop growth by destroying weeds very effectively even under situations where manual and mechanical means are not applicable. Chemicals can be easily used on closely planted crops where other methods cannot be used. Deep rooted, vegetatively propagated weeds can be controlled by using translocated herbicides. Herbicides are cheaper than manual weeding. Most of the time one application of the herbicide is enough to get good control and hence reduces the cost and drudgery in weed control. Selective weed killers are effective in controlling weeds with similar morphology to crop plants.

Herbicide works fast and therefore weeds can be removed quickly in critical situations. Non-selective herbicides help to wipe out any vegetation (whether grass or broadleaf weeds) present on the application site with a single spray. It is easy to clear areas where houses and roads are to be built.

Disadvantages of Herbicides

Technical knowhow is required to choose the herbicide for a particular crop

It is essential to know the time of application, safe dose and method of application. Over dose of herbicides can be harmful to the crop while under-dose will not give the desired effect. Residual effects of some herbicides may cause harm to the succeeding crop. Herbicide drifts may cause harm to nearby crops. Erratic and continuous application of herbicides creates toxicity problems to crop plants, residual effect on intercrops and succeeding crops in rotation and health hazards. Accumulation of herbicide residues can be observed in crop produce and ground water. Among the many effects of herbicides, the most alarming is the danger they pose to human health by direct exposure during the course of their application (i.e., while spraying pesticides), or the indirect affect caused by drift or residues on food.

Another important problem with broadcast applications is that they may be non-selective and hence are toxic to a wide variety of plant species, and not just the weeds.

Continuous use of the same herbicide in a location can lead to the development of herbicide resistant weeds. Build-up of herbicide-resistant biotypes is seen where the same chemical has been used repeatedly for several years. Triazine resistance has developed in most countries where these herbicides have been used for many seasons. The usefulness of a number of other herbicides, including paraquat, and sulfonylurea types has been affected by the development of resistant biotypes.

Effects of Herbicides on the Ecosystem

Herbicides are less toxic substances compared to insecticides, but they are also capable of polluting the environment. Herbicides vary in chemical composition and in the degree to which they pollute the environment. Generally, herbicides are sprayed rather early in the crop season, where the plant cover is little and so major portion of the chemical enters the soil. . Herbicides usually enter the soil environment in the order of 0.05-2.0 kg/ha. If herbicides are assumed to be evenly distributed in the top 15cm of soil, the resulting herbicide concentration will be approximately 0.03-1.0 mg/kg or even lower.

Herbicides are supposed to be the major ground water polluting pesticides. Soil and water are the two major recipients of all unwanted and toxic materials including herbicides from different sources. Certain herbicides like simazine and other triazines are highly persistent in soil and water. The main reason accounting for residues of highly persistent herbicides in ground and surface water is the widespread use of these chemicals at high doses. Ground water contamination due to herbicides have been reported all over the world. Ali and Jain (1998) reported that concentration of atrazine, alachlor, simazine, metolachlor and prometryn in ground water were much higher than the permissible health value (0.1 $\mu g\ L^{-1}$) prescribed by United States Environmental Protection Agency (Vettorazzi, 1979).

Heavy doses of herbicides belonging to the group sulfonyl ureas affect microbial population of the soil. Herbicides targeting amino acid synthesis in plants (e.g. glyphosate) may inhibit N_2 fixation in soil. Most herbicides are specifically plant poisons, and are not very toxic to animals. As these chemicals change the vegetation of treated sites, birds, mammals, insects, and other animals foraging on the flora of the location are also affected, hence herbicides are capable of changing the natural habitat.

Pollution Potential of Herbicides

Herbicides are applied either to the soil or plant foliage. Most of them are used pre-emergence i.e., they are applied to the soil before the emergence of weeds. For effective control of weeds soil applied herbicides must remain in soil in an active and available form until their mission is completed. But they should be degraded soon after the desired period. Herbicide persistence is important for effectiveness of the chemical but it also decides the pollution potential. Persistence of herbicide refers to the residence time of chemical in soil, before being completely removed by physical, chemical or biological degradation (Scheunert et al., 1993). Residence time of a particular herbicide in soil is determined by balance between adsorption to soil colloids, uptake by plants (residue), transformation or degradation processes and losses in liquid or gaseous form. This balance is partly controlled by the type of herbicide, its quantity and method of application and partly by soil characteristics such as texture and organic matter content e,g. 2,4-D, being an anion is little adsorbed by soil at normal pH of soil but because it degrades rapidly has a very low pollution potential. On the other hand, atrazine has low mobility in soil due to strong retention and causes ground water pollution problems in many countries due to its resistance to degradation (Kookana and Aylmore, 1993). Herbicide transformation in soil and the performance of soil applied herbicides against weeds are important aspects that affect safety of the crop and the environment.

References

Ali, Imram and C. K. Jain. 2001. Pollution potential of toxic metals in the Yamuna river at Delhi, India. J of Environmental Hydrology.Vol.9 paper 12

Basu, S. and Rao, Y.V.2020.Environmental effects and management strategies of the herbicides. International Journal of Bio-Resources & Stress management,11(6)508-535

György Matolcsy, Miklós Nádasy, Viktor Andriska.1988. Herbicides . Studies on Environmental Sciences Vol: 32, Page: 487-786 ISSN: 0166-1116,

https://www.sciencedirect.com/journal/science-of-the-total-environment/vol/132/issue/2

Kookana, R.S. and Aylmore, L.A.G.1993. Retention and release of diquat and paraquat herbicides in soils. Australian Journal of Soil Research. V; 31(1).97- 109

Scheunert, I., Mansour, M., Dorfler, U. and Schroll, R. 1993. Fate of pendimethalin, carbofuran and diazinon under abiotic and biotic conditions. Science of the total environment. Vol.132,2-3.361-369

Sharma, N., Yaduraju, N.T. and Rana, S.S. 2018. Herbicides vis a vis other pesticides for weed control in greengram.Indian Journal of Weed S cience,49 (3),252-255.

Vettorazzi, G. 1979. International Regulatory Aspects for Pesticide Chemicals, Volume 2. CRC Press, 216p

7

Herbicide Classification and Selectivity

A large number of herbicides are currently available in the market hence selection of suitable herbicide needs judicious understanding of the product and the requirement. When large number of chemical options are available it can make the process of selection a bit confusing. So it is important to know the crops in which a herbicide can be used, the weeds it will control, the appropriate rate, adjuvants included in the formulation. The different uses of the chemical and the different situations for which it will be useful. In addition to these it is important to understand the mode of action of the herbicide and also the mechanism of action of the chemical. Considering all these factors there are different methods by which the herbicides are classified.

A few important currently applicable classifications are given below.

I. Classification of Herbicides According to Chemical Composition

The chemistry of a compound determines its interaction with biological and physical systems such as plants, animals, soil, water and atmosphere. Grouping of herbicides having similarity in chemical structure is the basis for chemical classification and this is useful in characterising herbicides. Broadly they are categorized into two major group's viz., inorganic and organic herbicides.

Inorganic Herbicides

Inorganic herbicides contain no carbon atoms in their molecules. Some acids and salts which are toxic to plant species were recommended for weed control. These were the first chemicals used for weed control. They were used between1869 and 1930, before the introduction of organic herbicides. However, due to non-selective nature, high mammalian toxicity and extended persistence in soil these chemicals are no longer recommended for weed control.

a) Acids: Arsenic acid, arsenious acid, arsenic trioxide, sulphuric acid.

b) Salts: Borax, copper sulphate, ammonium sulphate, Na chlorate, Na arsenite, copper nitrate.

II. Organic Herbicides

Discovery of selective nature of organic compounds in controlling weed species was a breakthrough in herbicide chemistry. Since 1950 substantial number of herbicides of different chemical nature and with different modes of action were developed. From the chemistry point of view, herbicides were classified into different chemical groups, viz., phenoxy alkanoic acids, triazines, ureas, sulphonyl ureas etc. There are more than 30 classes of organic herbicides (Table 1). The classification of organic herbicides is a complex one and periodically modified as and when new herbicides are developed. Herbicide classification of the Weed Science Society of America (Retzinger and Mallory – Smith (1997) is followed here with few alterations. Only the important classes are elaborated here.

Table 1: Chemical classification of herbicides

Chemical group	Chemical sub group	Candidate herbicide
Acetamides	Acetamide	Napropamide
	Benzamides	Tebutam
	Oxyacetamides	Fluthiamid
	Sulfonamides	Perfluidone
	2-chloroacetanilides	Propanil, Butachor, Pretilachlor
	N-arylalaninanilides	Benzoylprop-ethyl
	nicotinanilides	diflufenican
Aliphatics(aldehyde)		Acrolein
Aliphatic acids	Acetic acid	MCA
	Propionic acid	Dalapon
Azoles	Benzothiadiazoles	Bentazon
	Isothiazoles	-
	Isoxazoles	Isoxaflutole
	Isoxazolidinones	Clomazone
	Oxadiazoles	Oxidiazon
	Oxadiazolidines	Methazole
	Pyrazoles	Pyrazolynate
	Tetrazolinones	
	Thiazoles	Benazolin
	Triazoles	Amitrole
	Triazolinones (Phenyl)	Carfendrazone
Benzoic acid		Dicamba
Benzonitriles/nitriles		Bromoxylin

Chemical group	Chemical sub group	Candidate herbicide
Benzylethers/Cineoles		Cinmethylin
Bipyridilium		Diquat, paraquat
Carbamates and	Carbamates	Aslum
Thiocarbamates	Dithio-carbamates	Metham
	N-phenyl carbamates/ carbanilates	Barban desmedipham
	Thiocarbamates	Thiobencarb,cycloate
Cyclohexanediones		Setoxydim, cycloxydim
Dinitroanilines		Butralin, trifluralins
Dinitrophenols		Dinosam, DNOC
Diphenylethers		Nitrofen, oxyfluorfen
Heterocyclics	Benzofurans	Etofumesate
	Phenylpyridazines	Pyridate
	Pyridazinones	Norflurazon
	Pyridine carboxylic acid	Picloram
	Pyridines	Thiazopyr
	Pyridinols/pyridinones	Fluridone
	Quinoline carboxylic acids	Quinclorac
	Thiadiazinones	Dazomet, DMTT
Imidazolinones	Imidazolinones	Imazethapyr
Imidazolidinones	Imidazolidinones	Buthidazole
Mercurics		PMA
Organic arsenicals	Dimethyl arsinic acid	Cacodylic acid(CA)
	Methyl arsenic acids	MSMA, DSMA
Organophosphorus	Glycine & phosphonates	Glyphosate
	Phosphinic acids	Glufosinate-AM
	Phosphoroamidates and phosphoroamidothioate	Anilofos DMPA
	Phosphorodithioate	Bensulide
Phenoxy(-aryloxy) Phenoxy propionates		Fenoxaprop-p-ethyl
Phenoxy alkanoic acids	Acetic acids	2,4-D. ,2,4,5-T, MCPA
	Butyric acid	2,4-DB, MCPB
	Propionic acid	Fenoprop, Silvex
Phenoxy ethyl sulphates/ phosphates		MCPES, 2,4-DEP
Phenoxyalkanoic acids		Fenac/Chlorfenac
Phenylureas		Diuron, Linuron
Phthalic & phthalamic acids	N-phenylphyhalimides	Flumioxazin
	Phthalic acids	Endothall

Chemical group	Chemical sub group	Candidate herbicide
	Phthalamic acids	Naptalam
Pyrazoliums		Difenzoquat
Pyrimidinyl-thiobenzoates		Bispyribac
Dicarboxamides		Flumioxazin
Sufonylureas	Pyrazolesulfonylureas	Halosulfuron-methyl
	Pyridylsulfonylureas	Nicosulfuron
	Sulfonylureas	Bensulfuron-methyl, Chlorimuron-ethyl
Triazines	Chloro-triazines	Atrazine
	Methylmercapto/methyl Thio-triazines	Ametryn, Simazine
	Methoxy-triazines	Simetone, terbumetone
	Triazinones	Metribuzin
Triazolopyrimidine		Diclosulam
Sulfonanilides		Penoxsulam
Uracils		Bromacil
unclassified		Bromobutide, flupoxam

1. Oils and Non-oils

Compounds containing carbon and hydrogen in their molecules are included in this category. The effectiveness of emulsified xylene-type aromatic solvents for control of submerged aquatic weeds in flowing water was discovered in 1948. Other chemicals under this category used as herbicides include, diesel oil, standard solvent, xylene-type, aromatic oils, polycyclic aromatic oils etc. These chemicals have been used to kill general weed growth on roadsides and railway road beds to prevent the spread of fires.

2. Aliphatic Compounds (Haloalkanoic acids)

Aliphatic compounds contain carbon, hydrogen and oxygen. The haloalkanoic acids are very active against grasses. Important herbicides of this class are dalapon, TCA, acrolein, glyphosate and methyl bromide. There are both contact and systemic herbicides. Sodium trichloro acetate (TCA) is a contact herbicide while dalapon and glyphosate are systemic herbicides.

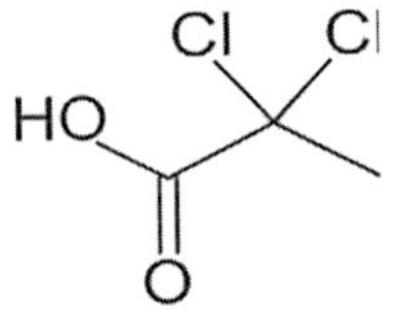

TCA(trichloroacetic acid)

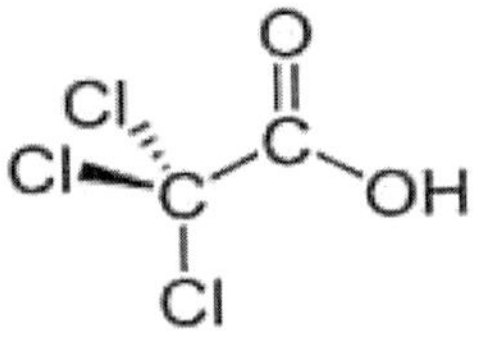

Dalapon(2,2-dichloropropionic acid)

3. Phenox yalkanoic Acids

Phenoxy alkanoic acid herbicides have phenoxy moiety with chloro, methyl or methoxy substitutions and a side chain that is based on acetic acid, propionic acid or butyric acid. Phenoxyalkanoic acids are divided into three groups viz., phenoxy acetics, phenoxy butyrics and phenoxy propionics.

Phenoxy Acetics

The most popular herbicide 2,4-D is coming under phenoxy acetics group. The advent of 2,4-D in 1944 resulted in dramatic growth of herbicide research and development of wide range of organic herbicides .

History of Development of 2,4-D

Foundations for the development of organic herbicides for weed control were laid out during the 1939-45 World War II period. Dramatic growth in the synthesis and introduction of organic herbicides in their present form is due to the discovery of hormone herbicides . In the 1920's it was found that plants produce a hormone IAA (Indole-3- acetic acid) which plays a major role in controlling their growth. It was during the early days of 1939-45 World War II that scientists realized the importance of this knowledge and its application in killing weeds. Synthetic compounds related to IAA were applied to a wide range of plants as rooting compounds by Zimmerman and Wilcox in 1935. Subsequent works showed that acetate side chain was essential for auxin behaviour and the chlorinated phenoxy compounds could act as analogues for indolyl compounds. Their lower toxicity and less polar nature led quickly to the successful demonstration of their herbicidal characteristics. Synthesis of 2,4-D and 2, 4, 5-T was first accomplished by Pokorny of C.B. Dolge company, Westport, Conn in 1941 (Pokorny, 1941). For this, equimolar quantities of 2,4- dichlorophenol and 2, 4, 5-trichlorophenol were mixed separately with monochloracetic acid and heated with a slight excess of sodium hydroxide together with water and evaporating almost to dryness. Both were white odourless crystals almost insoluble in water. Zimmerman and Hitchcock (1942) described the use of 2,4-D as a plant growth regulator but not as a herbicide. Marth and Mitchell in 1944 in USA reported that it killed dandelions and plantains in lawns. In the same year , Hamner and Tukey described field trials using 2,4-D as herbicide and they published a report on the selective herbicidal action of 2,4-D and 2, 4, 5-trichlorophenoxy acetic acid (2, 4, 5-T) on bindweed and on the superior effectiveness of 2, 4, 5-T as a brush killer. Another potent herbicide of phenoxyalkanoic acid group is MCPA ([4 - chloro - o-tolyloxy] acetic acid). Both MCPA and 2,4 D were quickly adopted , the former is preferred in United Kingdom and the latter in

United States of America. However, these two herbicides are not active against all dicotyledonous weeds. In 1944 itself, Amchem introduced 2,4,5-T (2,4,5-trichlorophenoxy acetic acid) as "Weedone" which resembles 2,4-D but much more effective for the control of many woody species.

Phenoxy alkanoic acids are formulated as sodium / potassium salts, amine salts and esters. The esters of phenoxyacetic acids are liquids having high lipoid solubility and more volatile than free acids and their salts (Rao, 1992). Esters are available as emulsifiable concentrate formulations. As volatility is a factor in reducing the effectiveness of a herbicide, it is recommended to incorporate such herbicides into the soil or use improved formulation. Sodium salt of 2,4-D, available in the form of dust or wettable powder is the most popular herbicide in India . It is highly soluble in water (4.5 g per 100 ml at 25^{0}C). Therefore, residues of 2,4-D can enter ponds and streams by direct application or drift, by inflow of herbicide previously deposited in dry stream beds, pond bottoms or irrigation channels; run off from soils; or by leaching through soil columns (Norris, 1981).

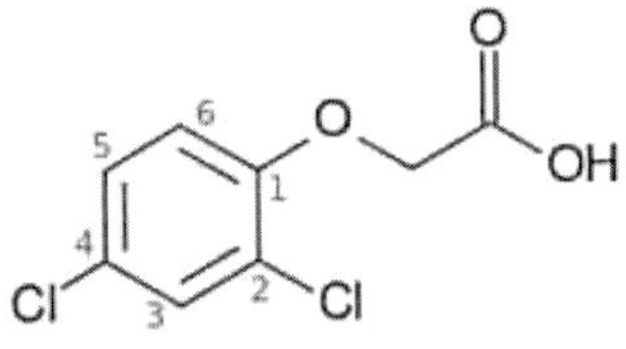

2,4-D

Amine salts are in the liquid form (soluble concentrate) and have equal solubility in water and lipid. Their persistence in soil is greater than sodium salts due to higher adsorption by clay particles and uptake by plant roots

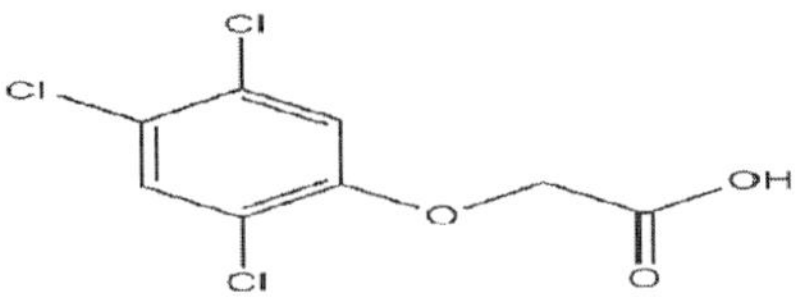

2,4,5 _T

Phenoxybutyrics

The first phenoxybutyric herbicide is MCPB ([4 - chloro - o-tolyloxy] acetic acid) and it was introduced by May and Baker in 1954 as "Tropotox". 2,4-DB(2,4 – dichlorophenoxy butyric acid) was introduced as " Embutox" by the same company in 1957. Both are formulated as alkali metal and amine salts and as esters which are applied for broad leaved weed control in pulse crops.

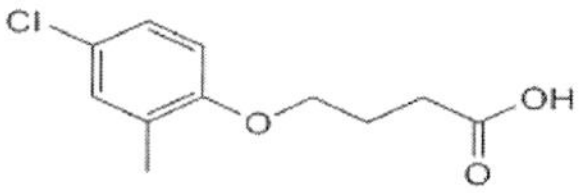

MCPA 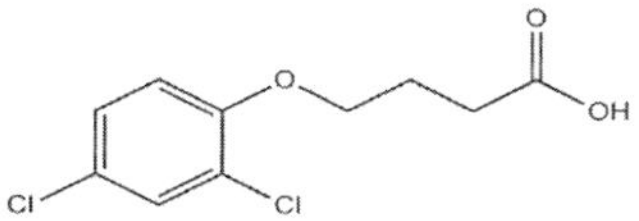 2,4-DB

Phenoxypropionics (Aryloxy phenoxy propionics)

Fluazifop-p-butyl(2,[4-(4-trifluoromethyl-2-pyridyloxy] phenoxy propionate), fenoxaprop –p- ethyl, quizalofop-p-ethyl, clodinafop-propargyl etc. are the important members of this group. These herbicides are called as " double ring graminicides"/"fop"herbicides. All these are post emergence herbicides having aryloxy phenoxy propionate structure which were developed for the selective control of annual and perennial grass weeds. They have little or no soil activity. Fenoxaprop p- ethyl and clodinafop-propargyl are called "wild oat herbicides" because they are highly effective in controlling wild oats in wheat and barley when they are applied at 28-35 days after sowing.

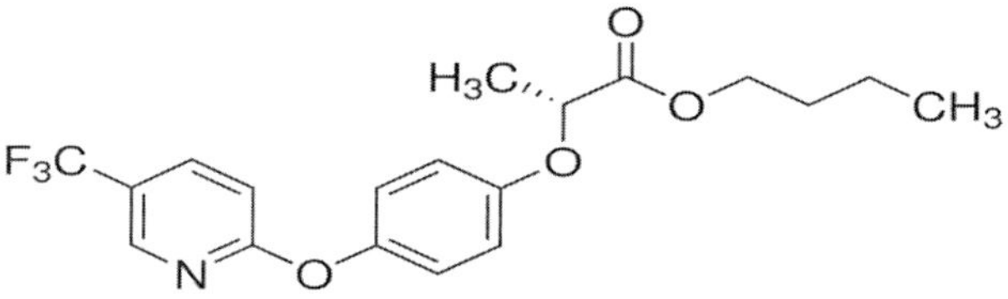

Fluazifop butyl

4. Amides (Acetamides and Acetanilides / phenyl amides)

Based on the valence and bonding order, nitrogen forms three bonds in its neutral state and maintains one pair of non-bonded electrons (NBEs). In amides one of the three bonds is a carbonyl carbon. Thus amides may be viewed as "acylated amines" or as derivatives of carboxylic acids in which the -OH of the acid has been replaced by -NR2 where R= H, alkyl, aryl, etc. Amides may be aliphatic (acetamides) or aromatic (anilides / benzamides / phenyl amides) based on the nature of the nitrogen substituents and overall structure. Aliphatic amides have simple hydrocarbon substituents (alkyl groups) while aromatic amides have at least an aromatic ring substituent.

Most of the acetamides (diphenamid, naptalam etc.) and anilides (alachlor, butachlor, metolachlor etc.) are selective herbicides and used as either pre-emergence or pre-plant herbicides in cereal crops such as rice, wheat, corn and barley. However, the anilide herbicide propanil (3,4-dichloropropionanilide) has contact activity and therefore used to control both broad leaved weeds and grasses mainly in rice fields.

Propanil

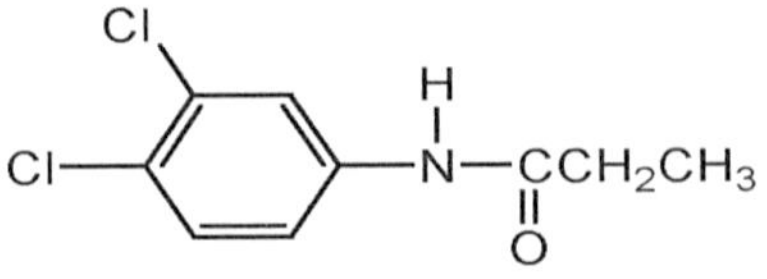

Butachlor (N-(Butoxymethyl)-2-chloro-N-(2,6-diethylphenyl)acetamide)

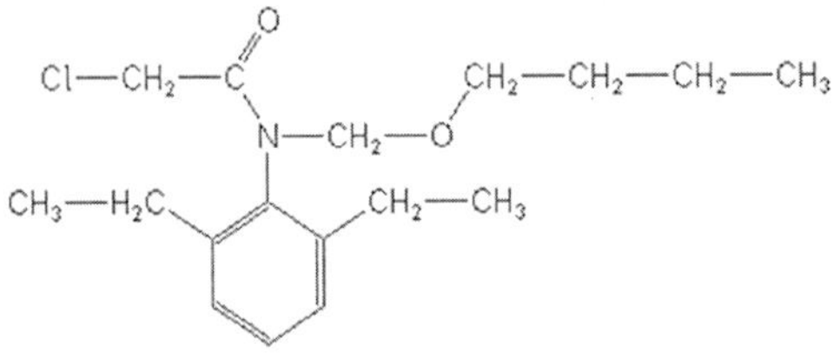

5. Benzoic acids (aromatic acids)

Herbicides under this category are more persistent in soil and are useful for the control of deep-rooted perennial weeds. The first herbicide based on aromatic acids was 2,3,6- TBA(2,3,6 –trichlorobenzoic acid) introduced by Du Pont in 1954 under the trade name "Trysben". Other herbicides coming under this group are chloramben, dicamba, fenac etc.

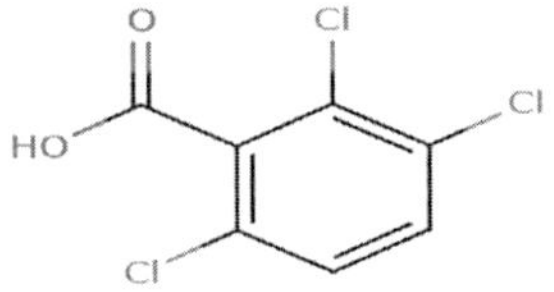

2,3,6 -TBA

6. Bipyridyliums

The name bipyridylium indicates the attachment of two pyridyl rings.

Most popular contact herbicides like paraquat (1,1′-dimethyl-4,4′-bipyridylium dichloride) and diquat (1,1'-ethylene -2,2'-bipyridylium dibromide) belong to this chemical class. They act as desiccants when in contact with aerial parts of plants. They are strongly adsorbed by soil and are inactivated. Paraquat is available under the trade name 'Gramoxone'. It destroys photosynthetic tissues and the first effects being noticeable after a few hours and kill is usually completed in 3-4 days. Diquat is mainly used to control aquatic weeds.

Picloram, triclopyr and 3,6 picolinic acid are the other herbicides belonging to this group.

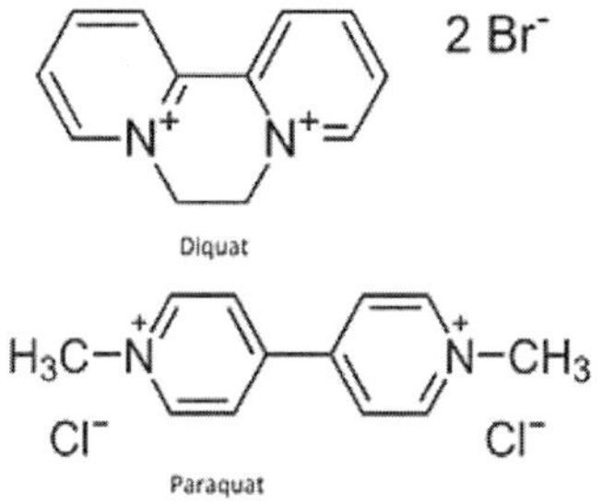

Diquat

Paraquat

7. Carbamates

The basic chemical structure of these category of herbicides is derived from carbamic acid. This class of herbicides rapidly degrade in soil except for chlorpropham which has high persistence in soil and can remain for one or two months. They are also moderately leached and can reach the ground water reserve. Most popular among this group of herbicides is chlorpropham [propan -2-yl N-(3-chlorophenyl) carbamate]. It is used for control of annual grasses and few dicots in lentils, pea, sugarbeet, garlic, onions, carrot and alfalfa etc. Another important herbicide under this group is Barban which is used against wild oats in barley, wheat etc.

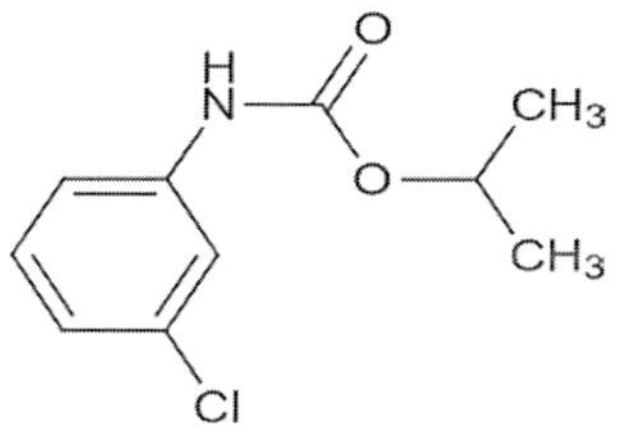

Chlorpropham [propan-2-yl *N*-(3-chlorophenyl) carbamate]

8. Thiocarbamates

They are derivatives of carbamic acid with one of the oxygen atoms replaced by a sulphur atom. They were first introduced by Stauffer chemical company in 1956. These herbicides are easily absorbed by soil and are not easily leached. As they are volatile in nature they are mixed with soil either by tillage or applied with irrigation water.They are selectively applied as pre plant or pre emergence for control of annual grasses and many dicot weeds e.g. butylate, di-allate, triallate, molinate, pebulate, vernolate, thiobencarb, asulam, cycolate. Thiobencarb(S-4-chlorobenzyl)-N, N-diethylthiocarbamate) was an important herbicide used for the control of weeds in rice showing very high selectivity between rice and barnyard grass.

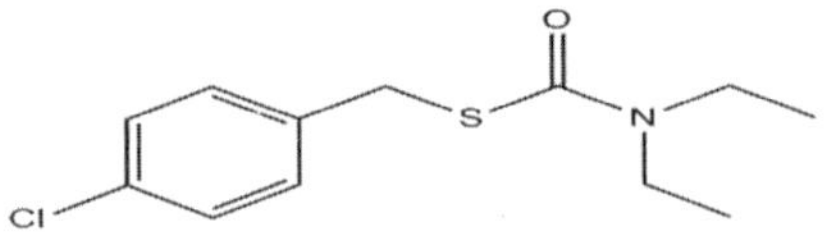

Thiobencarb

9. Dithiocarbamates

These compounds are amides formed from dithiocarbamic acid and they form stable metal complexes. Dithiocarbamate such as CDEC (2-chloroallyl diethyldithiocarbamate) is used as pre-emergence herbicide in vegetable crops. Metham sodium (sodium -N-methyl dithiocarbamate) acts as a soil fumigant, insecticide, herbicide and fungicide. It is generally used to control shallow perennial weeds.

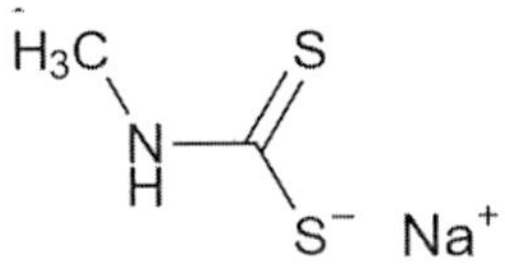

Metham sodium

10. Nitriles (formerly benzonitriles)

Nitriles have a benzene ring, containing CN or cyanide grouping. These are contact herbicides.They inhibit growth of sensitive species.They are tightly adsorbed to soil and have restricted leaching. Dichlobenil is applied to soil in orchard and also to control aquatic weeds while Bromoxynil and Ioxynil show high degree of selectivity to graminaceous crops and used against dicot weeds resistant to 2,4-D. Ioxynil has some systemic activity also.

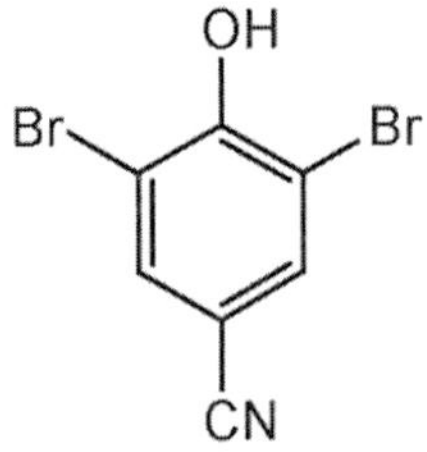

Bromoxynil

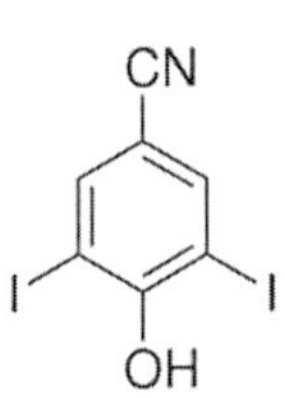

Ioxynil

11. Dinitroanilines (Toluidines)

Herbicides such as benefin, nitralin, trifluralin, butralin, dinitramine, fluchloralin, oxyzalin, penoxalin, pendimethalin etc. are included in this group. They are used exclusively as soil incorporated pre-emergence selective herbicides and are absorbed through root and affect shoot growth. When absorbed through shoot, they inhibit shoot growth directly and interfere with cell division

Pendimethalin [(3,4-dimethyl-2,6-dinitro-*N*-pentan-3-ylaniline) or (N - [1-ethylpropyl]-3,4-dimethyl-2,6-dinitrobenzenamine)] is used as pre and early post-emergence herbicide to control annual grasses and certain broad leaf weeds in cereals and horticultural crops.

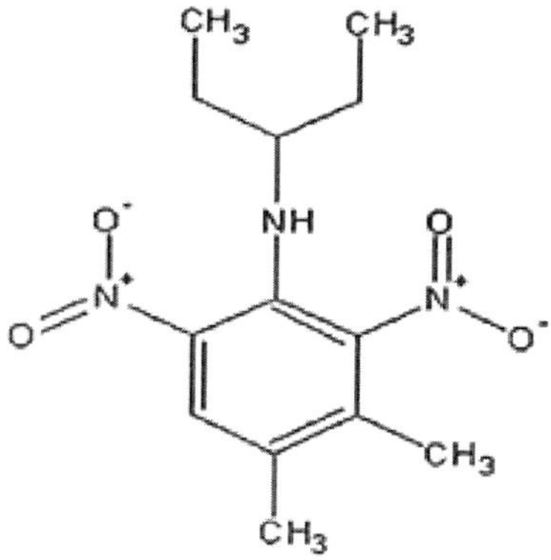

Pendimethalin

12. Triazines

Triazine compounds are nitrogen containing heterocycles which constitute 3 major classes (i) chloro triazines such as atrazine, simazine (ii) methoxytriazines such as prometone, atratone , metribuzin, and (iii) methylthiotriazine such as, ametryne, terbuteryne, propazine, prometryn. This class of chemicals was introduced by Geigy pesticide Laboratory, Switzerland in early 1950s.Triazine compounds inhibit photosynthesis by affecting the hill reaction.

Atrazine

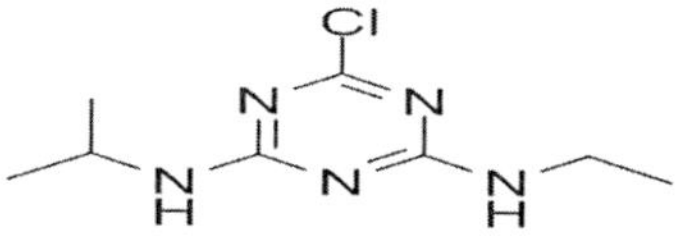

2-chloro-4-ethylamino-6-isopropylamino-1,3,5-triazine

13. Substituted Urea

Most of the urea herbicides such as monuron, diuron, flumeturon, methabenzthiazuron, chlorbromuron, chloroxuron, metoxuron are relatively non -selective and are mostly applied to soil. Some are active through foliage by adding surfactants.Two important herbicides under this group are (i) diuron [3-(3,4-dichlorophenyl)-1,1-dimethylurea] and isoproturon [3-(4-isopropylphenyl)-1,1-dimethylurea]. Diuron is widely used in pine apple as pre-emergence and post emergence herbicide for the control of annual and perennial broad leaf and grass weeds. It is also used for selective control of weeds including woody plants in cropped and noncropped areas. Isoproturon controls annual grasses and many broadleaf weeds in cereals (especially in wheat), sugarcane, cotton, citrus etc.

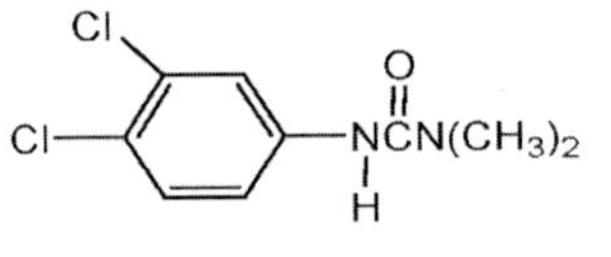

Diuron

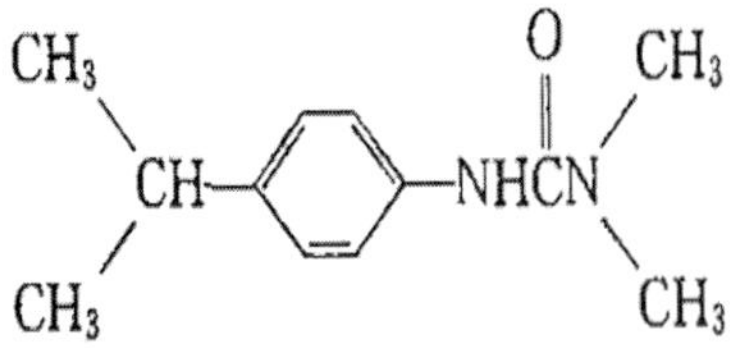

Isoproturon

14. Uracils

Uracil herbicides such as bromacil (5-bromo-3-butan-2-yl-6-methyl-1*H*-pyrimidine-2,4-dione) is used in combination with diuron or amitrole for total weed control on non-crop land. Bromacil inhibits hill reaction in photosynthesis. Other herbicides in this group are terbacil and lenacil.

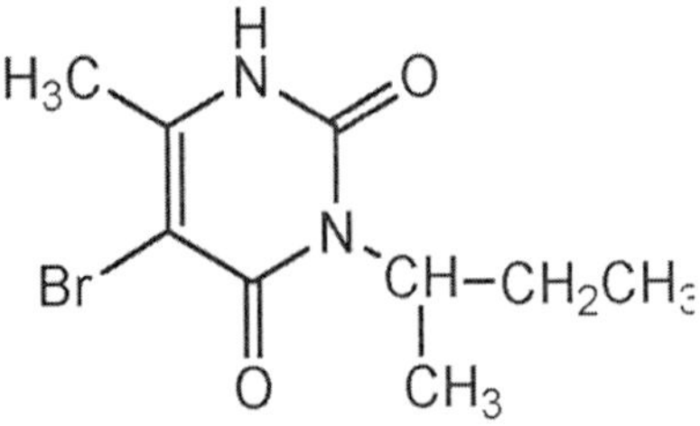

Bromacil

15. Diphenyl ethers

These herbicides have two phenyl rings tied together with oxygen. Important herbicides include nitrofen, oxyfluorfen (2-chloro-1-(3-ethoxy-4-nitrophenoxy)-4-(trifluoromethyl benzene), acifluorfen etc. They are used to control broad leaf and grass weeds in various crops. By contact action these chemicals interfere with photosynthesis by inhibiting an essential enzyme viz., protoporphyrinogen oxidase ("protox"in short) involved in chlorophyll biosynthetic pathway in plants. This leads to the death of plants.

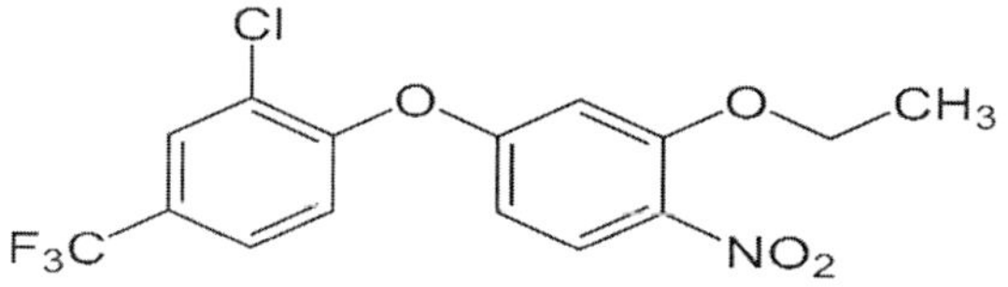

Cl, O, O, CH_3, F_3C, NO_2

Oxyfluorfen

It is having mainly pre-emergence activity both on grass and broad leaf weeds and used widely in cotton, soybean in dry areas. Wheat, corn and tomatoes are also unaffected by the application of oxyfluorfen. Pre-emergence application of oxyfluorfen is highly affective and pre-sowing application is recommended in Kerala for the control of weedy rice in paddy.

16. Organoarsenic Compounds

Major herbicides in this group such as MSMA (monosodium methyl arsonate), DSMA (disodium methyl arsonate) are the salts of methyl arsenic acid and dimethylarsenic acid (cacodylic acid) respectively. They are contact herbicides with some systemic property. Generally used for controlling weeds in cotton. Sodium salt of cacodilic acid is used as total weed killer in non-cropped areas and for killing unwanted trees by direct tree injection. Arsenate (AsV) is the oxidized form and occurs in well-aerated soils, whereas in chemically-reduced soil environments, arsenite (AsIII) is the prevalent As form. Although arsenite is more toxic than arsenate, arsenate can also have deleterious effects on humans, plants, and microorganisms. Arsenic-contaminated soils pose serious risk to human health.

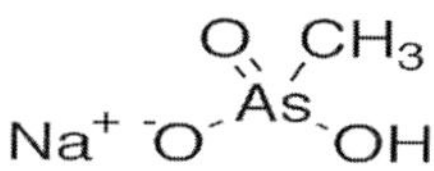

O, CH_3, As, Na^+ ^-O, OH

MSMA

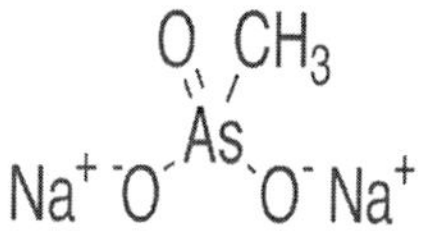

O, CH_3, As, Na^+ ^-O, O^- Na^+

DSMA

17. Phosphono Aminoacids (Organophosphorus Compounds)

This class of herbicides received considerable attention by the introduction of glyphosate (N-phosphonomethyl-glycine). The mechanism of action of organo phosphorus compounds is reported to be complex. Glyphosate is used in the form of isopropylamine salt for complete destruction of vegetation. Due to its systemic action it kills the root of perennial species and the rhizomes of grasses. They are considered to be nonselective foliar compounds with no soil activity. They penetrate rather slowly and hence receipt of rain shortly after application can reduce its effectiveness. Glufosinate is another organophosphorus compound which is used primarily for the control of grass weeds in orchards and vineyards as a post emergence contact herbicide.

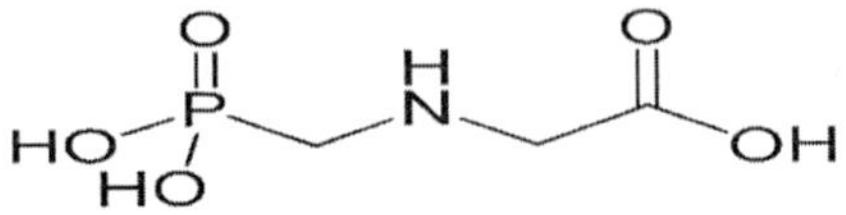

Glyphosate

18. Imidazolinones

These are a class of heterocyclic compounds derived from imidazoles. Most popular herbicides in this class are imazapyr, imazethapyr and imazaquin. They are used as broad spectrum herbicides for control of dicot weeds and grasses in perennial crops and legume crops. Imidazolines inhibit biosynthesis of branched chain amino acids like valine, leucine and isoleucine

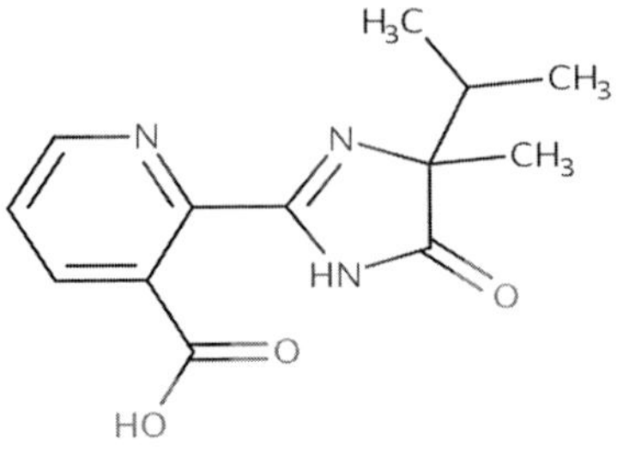

Imazapyr (2-[4,5-dihydro-4-methyl-4-(1-methylethyl)-5-oxo-1H-imidazol-2-yl]- 3-pyridinecarboxylic acid)

19. Sulphonylureas

Sulfonyl ureas are more potent herbicides than phenyl ureas. These chemicals inhibit the enzyme acetolactate synthase (ALS) resulting in impaired branched chain aminoacid synthesis. They are highly active compounds, selectively controlling dicotyledonous weeds in cereals at dose rates in grams rather than kg/ha and extremely popular worldwide because of their low mammalian toxicity and the above mentioned desirable characteristics. Metsulfuron methyl, chlorimuron ethyl, sulfometuron methyl, bensulfuron methyl, chlorsulfuron, triasulfuron etc. are the major herbicides of this group.

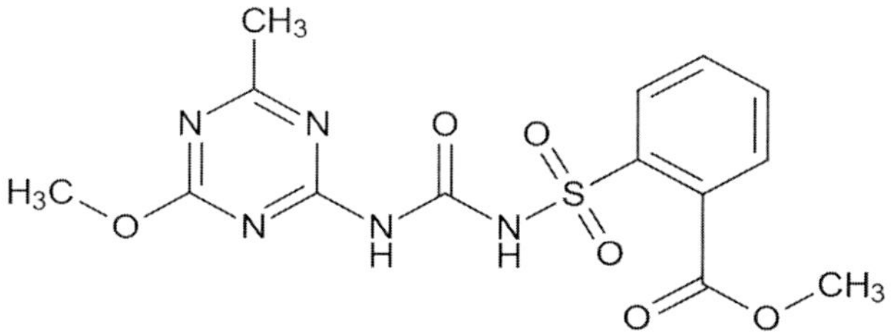

Metsulfuron methyl (methyl 2-[(4-methoxy-6-methyl-1, 3, 5-triazin-2-yl) carbamoylsulfamoyl]benzoate)

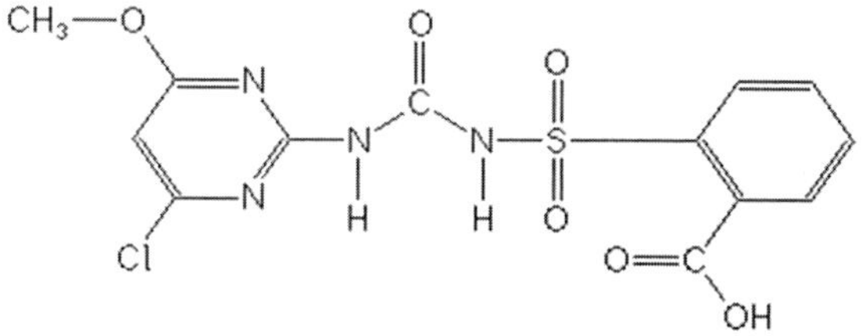

Chorimuron ethyl (ethyl 2-[(4-chloro-6-methoxypyrimidin-2-yl carbamoyl-sulfamoyl] benzoate)

20. Unclassified Compounds

These herbicides do not fit into any of the groups mentioned above. Herbicides like chloroflurenol, dymron/daimuron, flupoxam, cafenstrole, metam, pelargonic acid etc. are still not classified.

II. Classification Based on use of Herbicide

There are eight important points of differentiation for the classification of herbicides based on use viz., time of application, selectivity, spectrum of weed control, site of application, mode of action, window of application, residual action in soil and duration of weed control as depicted in the following scheme.

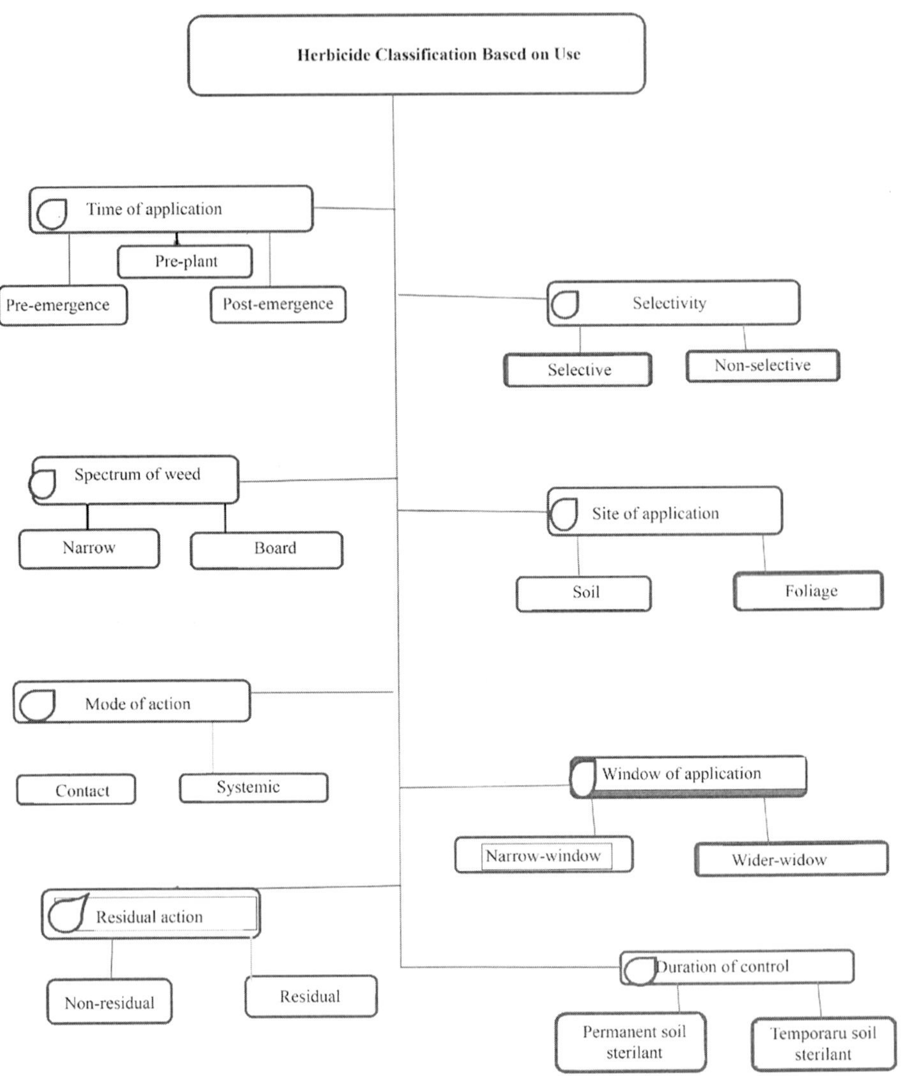

1. Classification of Herbicides Based on Time of Application

a. **Pre-Planting/ Pre-Sowing:** These herbicides are applied before a crop is planted and usually incorporated into the soil to reduce volatility and photo decomposition. E.g. trifluralin, fluchloralin

b. **Pre-emergence:** Pre-emergence herbicide is most effective when applied before the emergence of crop and weeds. The term may also be used for herbicides applied after the crop has emerged or established as in case of transplanted crop but before the emergence of weed. eg. simazine, atrazine, nitrofen, alachlor, butachlor

c. **Post-Emergence:** Post-Emergence herbicides are most effective when applied after the emergence of crop and weeds e.g. 2,4-D, MCPA, MCPB, propanil, glyphosate, paraquat etc.

2. Classification Based on Selectivity

Selective: Selective herbicides kill or retard the growth of some plants with little or no injury to other plants

Non-Selective: Non-selective herbicides are toxic to all the plants or kill all kinds of vegetation.

3. Classification Based on Spectrum of Weed Control

Narrow spectrum

Some herbicides are selective to dicot or monocot weeds. 2,4-D or Almix kills only dicot weeds while Cyhalofop butyl or Fernoxaprop-p-ethyl is used to control grasses. The spectrum of weeds controlled by these herbicides is very narrow .Some are even specific to certain weed species.

Broad spectrum

These are mostly non selective herbicides that kill all vegetation in an area. They do not discriminate between monocots or dicots and are used for clearing areas like road side, or in orchards and plantations. Glyphosate is one of the most popular broad spectrum herbicides.

4. Classification based on method of application

a. Foliage application: 2,4-D,Glyphosate, Cyhalofop butyl

b. Soil(Root) Application: butachlor, simazine, atrazine

   i. **Soil fumigants:** They usually function as a vapour or gas that diffuse through the soil and have relatively short life in the soil. Cyanamide, methyl bromide, carbon disulphide, trifluralin.

   ii. **Soil sterilants:** Any chemical which prevent the growth of green plants when present in the soil is considered as soil sterilants. TCA, sodium chloride, diuron, monuron, atrazine, fenac.

5. Classification based on mode of action

1. **Contact herbicides:** These herbicides kill only those plants or retards the growth of those plants which comes in direct contact. E.g. glufosinate ammonium paraquat

2. **Translocated herbicides:** Herbicides which are absorbed by one part of the plant and exert toxic action on other parts .These are also known as systemic herbicides. These absorbed chemicals upset the plant growth and metabolic processes. 2, 4-D, MCPB, MCPA, Glyphosate.

5. Classification of herbicides based on window of application

Narrow window

Narrow window herbicides have length/range of time of application highly reduced /limited to just before sowing or from sowing to before emergence of crop or at a specific time after crop emergence as post emergence e.g. trifluralin, fluchloralin, pendimethalin etc.

Broad window

Wider window herbicides as desired should have length / range of time span of application wider, e.g. from before sowing or after sowing to longer period of crop growth may be upto 40 days after emergence of crop. E.g. atrazine, butachlor

7. Classification of herbicides on the basis of residual effect

a) **Short persistent herbicides:** Residual effect remains in the soil upto a week. E.g. Paraquat, diaquat, amitrole, DSMA, DNBP.

b) **Medium persistent herbicides:** Residual effect remains in the soil for upto 2 to 6 weeks. Butachlor, oxyfluorfen, metolachlor, pretilachlor

c) **Very long persistent herbicides:** Residual effect remains in the soil for few months or even years.eg. Atrazine, simazine, diuron etc.

Table 2: Persistence of common herbicides in soil

Less than 3 months	3-6 months	More than 6 months
Alachlor	**Chlorbromuron**	Atrazine
Aminotriazole	Diallate	Bromacil
Anilofos	**Dinitramine**	Dichlobenil
Butachlor	**Isoproturon**	Diuron
Chlorpropham		
Cynazine		
Dalapon		
Fluchloralin		
Glyphosate	Linuron	
Metolachlor	Oxyfluorfen	
Metoxuron	Methabenzthiazuran	

Less than 3 months	3-6 months	More than 6 months
Metribuzin	Metobromuron	
Oxadiazon	Propyzamide	
Pretilachlor	Pyrazone	
Prometryn		
Propachlor		
Propanil		
Propham		
Sulfometuron-methyl		
Sulfosulfuron		
Terbutryn		
2,4-D		

Source: Sankaran et al., 1993

Mode of Action of Herbicides

Mode of action describes the way in which the herbicide controls the weed. It explains the biological process or enzyme in the plant that is disrupted by the herbicide (site of action) and how it affects normal growth and development. This is an important criterion in herbicide selection. Continuous use of a herbicide with the same mode of action in a location can lead to heavy selection pressure on a weed population and may eventually select for resistant individuals which may multiply and dominate the location. To prevent herbicide resistance in weed species it is important to rotate the herbicide applied and it should be seen that herbicides with the same mode of action should not be included in the rotation.

III. Classification of Herbicides by Primary Site of Action/mode of action

An understanding of the site at which herbicides act within plants helps to develop strategies that are able to reduce selection pressure for herbicide resistant weed biotypes.

Retzinger and Mallory – Smith (1997) developed a Weed Science Society of America (WSSA) - approved classification of herbicides by site of action. The international Herbicide Resistance Action Committee (HRAC) published a similar classification system. However, that system used letters instead of numbers for group designations (Schmidt, 1998). While minor differences exist between the two schemes, they both convey the same information. Classification of herbicides by site of action published by Retzinger and Mallory – Smith in 1997 was revised in 2003. The classification system uses a numbering system for herbicide's site of action, chemical family and common name (Mallory Smith and Retzinger, 2003). Regulatory agencies in the United States and Canada have published labelling guidelines based on this classification to aid in herbicide resistance management

This classification system is highly useful in addressing target site resistance and target site cross resistance. When a mutation occurs in the target site, target site resistance occurs which prevents the herbicide from binding. Target site cross resistance is the resistance in a biotype to herbicides from different chemical classes but with the same target site. This classification system is conservative in that all inhibitors of a target site come under one group. (Table 3).

Table 3: Classification of Herbicides According to Primary Site of Action Chemical Family and Mode of Action

Mode of Action	Site of Action Group	Site of Action	Chemical Family	Common Name	Trade Name
Lipid Synthesis Inhibitors	1 (A)	Inhibitors of acetyl CoA carboxylase	Aryloxyphenoxy propionate	Clodinafop Cyhalofop butyl Diclofop Fenoxaprop Fluazifopp Haloxyfop Propaquizafop Quizalofop-p	Discover(US)Horizon(Cada) Clincher Holeon, Whip, Acclaim, Fusilade 2000, Fusilade DX Assure, Targa super
			cyclohexanedione	Alloxydim Butroxydim Clethodim Cycloxydim Sethoxydim, Tralkoxydim	 Select, Prism, Poast, Poast Plus Achieve
Amino acid synthesis inhibitors	2 (B)	Inhibitors of acetolactate synthase (ALS). also called acetohydroxyacidsynthase (AHAS)	Imidazolinone	Imazamethabenz Imazamox Imazapic Imazapyr Imazaquin Imazethapyr	Assert Beyond, Raptor Cadre, Plateau Arsenal Scepter, Image Pursuit
			Pyrimidinylthio-benzoate	Bispyribac-sodium Pyrithiobac Pyribenzoxim	Staple
			Sulfonylamino-carbonyltriazolinone	Flucarbazonesodium, Propoxycarbazone	Everest Olympus, Attribute

Mode of Action	Site of Action Group	Site of Action	Chemical Family	Common Name	Trade Name
			Sulfonylurea	Amidosulfuron	
				Azimsulfuron	
				Bensulfuron	Londax
				Chlorimuron	Classic
				Chlorsulfuron	Glean, Telar
				Cinosulfuron	
				Cyclosulfamuron	
				Ethametsulfuron	Muster
				Ethoxysulfuron	Sunrise
				Flazasulfuron	
				Flupyrsulfuron-methylsodium	
				Foramsulfuron	
				Halosulfuron	Permit, Battalion
				Iodosulfuron	
				Mesosulfuron	Osprey
				Metsulfuron	Ally, Escort
				Nicosulfuron	Accent
				Primisulfuron	Beacon
				Prosulfuron	Peak, Exceed, Saathi
				Pyrazosulfuron-ethyl	
				Rimsulfuron	Titus
				Sulfometuron	Oust
				Sulfosulfuron	Maverick , Maverick pro
				Thifensulfuron,	Pinnaacle, Harmony
				Triasulfuron	Amber, Logran
				Tribenuron	Express
				Trifloxysulfuron-	Enfield
				sodium Triflusulfuron	Debut, Safari, Upbeet

Mode of Action	Site of Action Group	Site of Action	Chemical Family	Common Name	Trade Name
			Triazolopyrimidine	Cloransulam-methyl Diclosulam Florasulam Flumetsulam	First rate Strongarm Primus Broadstrike
Seedling root growth inhibitors	3(KI)	Inhibitors of microtubule assembly	Dinitroaniline	Benefin Ethalfluralin Oryzalin Pendimethalin Prodiamine Trifluralin	Balan, Sonlan Surflan prowl, Stomp Barricade Treflan,
			Pyridine	Dithiopyr Thiazopyr	Dimension Visor
			None	DCPA	Dacthal
Growth regulators	4	Synthetic auxins	Phenoxyalkanoic acids	2,4 –D 2,4 –DB Dichlorprop,2,4-DP MCPA MCPB Mecoprop, PP	Various Various Various Various Various Various
			Benzoic acid	Dicamba	Various
			Carboxylic acid	Clopyralid Fluroxypyr Picloram Triclopyr	Reclaim,Stinger, Various Starane Tordon, Grazon, various Garlon, various
			Quinoline carboxylic acid	Quinclorac (dicots)	Drive, Facet, Paramount

Mode of Action	Site of Action Group	Site of Action	Chemical Family	Common Name	Trade Name
Photosynthesis inhibitors	5(CI)	Inhibitors of photosynthesis at photosystem II siteA	Phenyl- Carbamate	Desmendipham Phenmedipham	Betanex Spin- aid
			Pyridazinone	Pyrazon	Pyramin
			Triazine	Ametryn Atrazine Cyanazine Desmetryn Prometon Prometryn Propazine Simazine Simetryn Terbumeton Terbuthylazine Trietazine	Evik AAtrex, various Bladex Pramitol Caparol, various Milo Pro Princep,various
			Triazinone	Hexazinone Metamitron Metribuzin	Velpar Sencor, Lexone
			Triazolinone Uracil	Amicarbazone Bromacil Terbacil	Bay MKH 3586,Bay 314666 Hyvar X Sinbar
	6(C3)	Inhibitors of photosynthesis at photosystem II site B	Benzothiadiazole	Bentazon	Basagran, Various

Mode of Action	Site of Action Group	Site of Action	Chemical Family	Common Name	Trade Name
	7(C2)		Nitrile	Bromoxynil Ioxynil	Butril, Brominal, Various Various
			Phenyl- Pyridazine	Pyridate	Lentagran, Tough
		Inhibitors of photosynthesis at photosystem II site A: different binding behavior from group 5	Amide	Propanil	Propanil, Stam, Various
			Urea	Chlorotoluron Dimefuron Diuron FluoMeturon Isoproturon Linuron Methibenzuron Metoxuron Monolinuron Siduron Tebuthiuron	 Karmex, Direx Cotoran, Meturon Lorox, Linex Tupersan Spike, Preflan

Mode of Action	Site of Action Group	Site of Action	Chemical Family	Common Name	Trade Name
Seedling shoot growth inhibitors	8(N)	Inhibition of lipid synthesis; not ACCase inhibition	Thiocarbamate	Butylate Cycloate EPTC Esprocarb Molinate Pebulate Prosulfocarb Thiobencarb Triallate Vernolate	Genate+ Ro-Neet Eptam, Eradicane Ordram Tillam, Edge Boxer, Deli, Arcade Bolero, Saturn Avadex, Fargo Vernam
			None	Bensulide	Prefar, Betasan, various
Amino acid synthesis inhibitors	9(G)	Inhibitors of 5-enolypyruvyl-shikimate-3-phosphate synthase (EPSP)	None	Glyphosate	Round up, Touchdown, Various
Nitrogen metabolism	10(H)	Inhibitor of glutamine synthase	None	Glufosinate	Ignite, Liberty, Rely, Basta, Sweep power
Pigment inhibitors	11(F3)	Inhibitors of carotenoid biosynthesis (unknown Target)	Triazole	Amitrole Aclonifen	Amitrol T. Amizol
	12(F1)	Inhibitors of phytoene deasaturase (PDS)	Pyridazinone Pyridinecarboxamide Other	Norflurazon Diflufenican Picolinafen Beflubtamid Fluridone Fluorochloridone Flutamone	Zorial, Evital, Solicam UBH-820 Sonar

Mode of Action	Site of Action Group	Site of Action	Chemical Family	Common Name	Trade Name
Diterpene synthesis inhibition	13(F4)	Inhibitors of 1-deoxy-D-xylulose 5-phoshate synthatase (DOXP Synthase)	Isoxazolidinone	Clomazone	Command
Cell membrane disrupters (PPO Inhibitors)	14(E)	Inhibitors of protoporphyrinogen oxidase (Protox)	Diphenylether	Acifluorfen Bifenox Fomesafen Fluoroglycofen Lactofen Oxyfluorfen	Blazer Reflex, Flexstar Cobra Goal
			N-phenylphthalimide	CGA-248757 Flumiclorac Flumioxazin	Action Resource Valor
			Oxadiazole	Oxadiazone Oxadiargyl	Ronstar
			Phenylpyrazole	Pyraflufen-ethyl	
			Pyrimidindione	Butafenacil	Inspire, CGA-276854
			Thiadiazole	Fluthiacet-methyl	KIH-9201, CGA-248757
			Triazinone	Carfentrazone-ethyl Sulfentrazone	Aim, Teamwork, Affinity Authority
			Triazolone	Azafenidin	Milestone
			Other	Flufenpyr-ethyl	V-3153

Mode of Action	Site of Action Group	Site of Action	Chemical Family	Common Name	Trade Name
Seedling shoot growth inhibitors	15(K3)	Inhibitors of Synthesis of very long –chain fatty acids	Acetamide	Napropamide	Devrinol
			Chloroacetamide	Acetochlor Alachlor Butachlor Dimethenamid Metolachlor Metazachlor Pretilachlor Propachlor Thenylchlor	Harness, Surpass, TopNotch Lasso Machete Frontier, Outlook Dual Rifit Ramord
			Oxyacetamide	Mefenacet Flufenacet	 Define
			Tetrazolinone	Fentrazamide	
			Other	Anilofos	
	16(N)	Unknown	Benzofuran	Ethofumesate	Nortron
	17(Z)	Unknown	Organoarsenical	DSMA MSMA	Various Various
Amino acid synthesis/ cell dvision inhibitors	18(I)	Inhibitors of 7,8-dihydro-pteroate synthetase (DHP)	Carbamate	Asulam	Asulox
Auxin transport inhibitors	19(P)	Inhibitors of Indoleacetic acid transport	Phthalamate Semicarbazone	Naptalam Diflufenzopyr	Alanap

Mode of Action	Site of Action Group	Site of Action	Chemical Family	Common Name	Trade Name
Cell wall biosynthesis inhibitors	20 (L)	Inhibitor of Cell wall Synthesis Site A	Nitrile	Dichlobenil	Casoron, Norosac, Various
-Do-	21(L)	Inhibitor of cell wall synthesis site B	Benzamide	Isoxaben	Gallery
Cell membrane disrupters	22(D)	Photosystem I electron diverters	Bipyridyium	Diquat Paraquat	Reglone, Diquat Gramoxone, Cyclone, Starfire
Cell division inhibitors	23(k2)	Inhibitor of mitosis	Carbanilate	Carbetamide	
Membrane disrupters	24 (M)	Membrane disruptors (uncouplers)	Dinitrophenol	Dinoterb	
	25 (Z)	Unknown	Arylaminopropionic acid	Flamprop	Mataven
Cell wall biosynthesis inhibitors	26(Z)	Inhibition of cell wall synthesis	Various	Cinmethylin Dazomet Difenzoquat Fosamine Metham Pelargonic acid Quinclorac (monocots)	 Avenge Scythe Facet,Drive
Pigment inhibitors	27(F2)	Inhibitors of 4-hydroxyhenyl-pyruvate-dioxygenase(4-HPPD)	Isoxazole Pyrazole Triketone	Isoxaflutole Benzofenap Pyrazolynate Pyrazoxyfen Mesotrione Sulcotrione	Balance Callisto

Herbicide Selectivity

Herbicide selectivity has been a boon to modern agriculture. The realization that herbicides like 2,4-D can selectively kill broad leaved weeds in crop production systems has helped to reduce the drudgery of hand weeding in crop production

According to Jones et al. (1963): "A selective herbicide is a chemical used in such a manner that it will kill weeds in a growing crop without damaging the crop, or will eliminate only the unwanted vegetation." Selectivity is a fickle, dynamic process. It is the result of complex interaction between plant, herbicide and environment. Selectivity is a relative property of herbicide and is dependent on time, method, rate and formulation, plant species, stage of growth of both crop and weed and environmental condition.

Kinds of Selectivity

Herbicide selectivity is derived from differences among plants in the interception and uptake of herbicides, differences in metabolic path-ways and rates, in target-site sensitivity and in tolerating product phytotoxicity (Devine et al., 1993). Herbicide selectivity can be the result of both biotic factors of the plant and chemical factors of the formulations. Hence it is categorized as physical selectivity, chemical selectivity, biological and chronological selectivity.

Physical Selectivity

Here the chemical applied may not be selective but by manipulating some physical conditions such as depth of sowing, protection is given to the crop plant seeds from the pre-emergence herbicide. In some cases activated carbon is spread over the rows where the seeds are sown. This will absorb the applied herbicide and protect the crop seed from injury. Non-selective herbicides can be used for spot application or contact application by using spray hood or by wick applicators.

Chemical Selectivity

Molecular makeup of the herbicides especially in the functional groups or side chains of phenyl or benzene ring is an important factor that contributes to selectivity. Both chlortoluron and diuron are herbicides belonging to the phenyl urea group. In the former the side chain on the 4th carbon position of the phenyl ring is a CH group while in the latter it is a Cl group. Due to this difference chlortoluron is selective to wheat while diuron is selective to cotton.

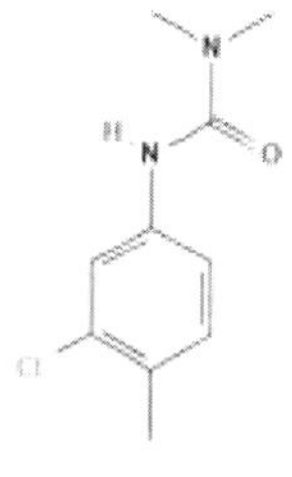

Chlorotoluron | C10H13ClN2O

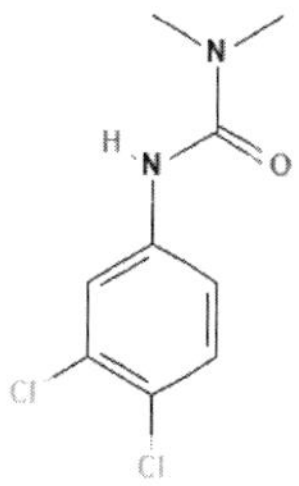

-Diuron C9H10Cl2N2O

Biological/Biochemical Selectivity

Genetic aspects of the plants may contribute to selectivity. This may be due to variation in the metabolic factors that can change molecular structure of the applied herbicide and deactivate the active ingredient in the herbicide formulation. The processes include enzymatic beta-oxidation or conjugation. In conjugation there is coupling of intact herbicide molecule with some plant cell constituents which will deactivate the herbicide.

Chronological Selectivity

Chronological selectivity is achieved by manipulating the time of application of herbicide. Some herbicides are recommended as both pre-pant, pre-emergence and post- emergence herbicide. In some method of application they show selectivity. Pendimethalin is selective to most crops when applied as pre- emergence but when applied as post emergence herbicide it does not show this character.

Factors Affecting Selectivity

Selectivity is the result of complex interaction between plant, herbicide and environment.

Plant Factor

Morphology of the plant is a major factor that decides selectivity. These include leaf and foliage factors such as nature and orientation of leaf, pubescence and waxiness, roughness and corrugation of leaf, composition and thickness of cuticle, Root system, age and stage of the plants.

Physiological factors of the plant are the differences in absorption, translocation and metabolism of herbicide in the plant system. 2, 4-D is selective to winter grains at first 3-6 leaf stage and dough stage. In other stages malformations like onion leafing, missing spikelet ear and tail ear occurs. In barley, malting quality is decreased by applying 2,4-D at susceptible stages.

Herbicide Factors

Chemical structure and molecular make-up of herbicide formulation, time of herbicide application, dosages of herbicide adjuvants and safeners, antidotes or crop protectants all are factors that can contribute to selectivity of herbicides.

Environmental Factors

Plant growth and development is dependent on climatic factors like light, temperature, humidity and rainfall. Temperature increases uptake translocation and metabolism in plants. This is true for soil absorbed herbicide. Atrazine is selective to maize but under cool climate it shows phytotoxicity symptoms on maize due to lower levels of herbicide metabolism. Under higher light conditions penetration of herbicides may be higher and phytotoxicity symptoms are observed for contact herbicides like diquat. Rains may wash away the herbicide and reduce their effectiveness. High humidity increases the penetration of herbicides and also their activity.

Soil factors

Higher organic content in soil will lead to higher adsorption of herbicides. So more chemical will be needed to achieve the same efficiency. Light textured soils will contribute to loss by leaching and the leached herbicide will reach the root zones of crop plant which can cause phytotoxicity symptoms in crop plants. In warmer soils, metabolism of herbicide increases than in cold soil. Persistence of herbicide and residual activity depends on the microorganisms in the soil. pH of the soil decides the availability of most of the herbicides and their selectivity.

References

Das, T.K. and Nath,C.P. 2015. Herbicide: History, classification, activity and selectivity. In Yaduraju, N. T. Sharma, A. R. and Das, T. K. (Eds.).2016. Weed Science and Management.402p.Indian Society of Weed Science, Jabalpur and Indian Society of Agronomy, New Delhi

Devine,M., Duke,S.O, Fedtke,C.1993. Physiology of herbicide action.Englewood Cliffs:Pentice Hall. 441p.

Hamner, C.L. and Tukey, H.B. 1944. The herbicidal action of 2,4-D and 2, 4-5-T on bindweed. Science 100: 154-155

Jones,G.E., Anderson,G.W.,Way/V!ell,C.G.Switzer,C.M and Bibbey,R.O.1963.Guide chemical weed control. Ontario Dept.Agric.Pub.75.pp55.

Mallory –Smith C.A. and Retzinger, E. J. 2003. Classification of herbicides by site of action for weed resistance management strategies. Weed Technol. 17:605-619

Norris, LA. 1981. “The Movement, Persistence, and Fate of the Phenoxy Herbicides andTCDD in the Forest.” Residue Reviews, 80, 65-135.

Pokorny, R. 1941. Some chlorophenoxyacetic acids. J. Am. Chem. Soc. 63: 1768

Retzinger, E. J. and Mallory – Smith. 1997. Classification of herbicides by site of action for weed resistance management strategies. Weed Technol. 11:384-393

Sankaran, S. Jayakumar, R. and Kempuchetty, N. 1993. Herbicide residues, Gandhi book house, Coimbatore, 251p.

Schmidt, R.R. 1998. Classification of herbicides according to mode of action. Leverkusen, Germany: Bayer Ag. 8p.

WSSA, 1994. Herbicide Handbook, Weed Science Society of America: Champaign

Zimmerman, P.W. and Hitchcockm A. E.1942. Substituted phenoxy and benzoic acid growth substances and the relation of structure to physiological activity. Contributionof Boyce Thompson Institute 12:321-344.

Zimmerman, P.W. and Wilcoxin, F.L. 1935. Several chemical growth substances which cause initiation of roots and other responses in plants. Contrib. Boyce Thompson Inst. 7: 209-229

8

Adjuvants

Effectiveness of spray mixtures can be increased by addition of adjuvants. Weed Science Society of America (WSSA1994) have defined adjuvant as a material in a herbicide formulation or added to the spray tank to modify herbicidal activity or application characteristics. Adjuvants enhance activity of a herbicide in different ways viz., wetting, spreading, deposit building, emulsifying or deflocculating the active ingredient. In many cases cost of adjuvant will be less than that of active ingredient. In such cases it helps to reduce the cost of the final product. Moreover adjuvants may have more than one mode of action and so improves effectiveness of the herbicide.

Effectiveness of any pesticide/herbicide depends on the penetration of the applied chemical into the plant system. As the leaf exteriors may contain waxes, cutin, pectin and cellulose, entry into the plant system may be inhibited. In many cases adjuvants help to overcome these barriers and allow uniform deposit and penetration of the applied chemical.

In some cases adjuvants are applied to stabilize products and to make them effective in the spray tank.

Classification of Adjuvants

There are two important classes of adjuvants viz., activator adjuvants and spray modifiers/ utility modifers which are used with herbicides. .

Activator Adjuvants

Activator adjuvants are of different types viz., surfactants, wetting agents, penetrants, oils and colorants. Among these, surfactants are key adjuvants used in most of the herbicide formulations.

Surfactants

Surfactants are surface active agents they reduce the interfacial tension of two boundary surfaces and help the spray solution to spread and cover surfaces more effectively. There are four major types of surfactants used in the herbicide formulations.(i) Anionic surfactants are the alkaline metal salts of organic acids (carboxylates, sulphonates , sulphates or phosphates) and are usually used with acids or salts. (ii) Cationic surfactants contain amino or quarternary ammonium unit which form cations in aqueous phase. (iii) Non ionic surfactants are neutral molecules and are generally compatible with most herbicides. The nonionic surfactant poly ethoxylated tallow amine (POEA) is used in many glyphosate formulations viz., Roundup , Glycel etc. Amphoteric surfactants have acidic as well as basic hydrophilic heads in the same molecule and produce anions and cations according to the type of environment. Alkylamphoacetates, alkylamphopropionates and alkyliminopropionates are the major class of chemicals coming under this group. Lecithin contained in egg yolks possesses both acidic and alkaline properties and is a type of phospholipid type amphoteric surfactant. Another source of lecithin is soybean

Organosilicone compounds are the most powerful nonionic surfactants and are used at much lower dose than other nonionic surfactants. Special properties of these compounds include extreme spreading behavior and exceptionally low surface tensions of their aqueous solutions and the ability to penetrate into the foliage via stomata. . This group of surfactants enabled the herbicide glyphosate to be used as a cost-effective alternative to 2,4,5-T for the control of forest scrub weeds. The first organosilicone adjuvant was commercialised in 1985 with the brand name Silwett-77

Wetting Agents/ Humectants

These compounds retain moisture of deposited herbicide and prevents the herbicide from crystallizing on the leaf surface, thus improving adsorption into the leaf.

Penetrants

Oil soluble penetrating agents increases the movement of the herbicide into and eventually through the cuticle so that the herbicide can be absorbed into the outer layer of cells

Oils

Oils improve the retention time of a solution on leaves, thus helps to increase the herbicide absorption. Crop oils, crop oil concentrates, vegetable oils etc. keep the leaf surface moist longer than water and enhance the activity of many herbicides. Methylated seed oils are vegetable oils mainly from oil seed rape or sunflower esterified with methanol to get methyl esters.

Colorants

Addition of colorants to herbicide helps to reduce quantities applied by allowing greater precision and accuracy. Colorant may be a dye or a pigment or a mixture of these two. A suitable dye marker allows spray patterns to be easily identified, giving early indications of drift to nontarget areas. Use of colorants reduce the skips and double application.

Spray Modifiers / Utility Adjuvants

Spray modifiers include stickers/ spreaders, drift inhibitors/thickening agents, compatibility agents, pH buffers, defoamers etc. which on addition to the formulation can widen the range of conditions under which a particular herbicide is useful.

Spreader Sticker

Sticker is a glue that makes the herbicide spray stick to the leaves. Natural gums, poly ethylene polysulphides, poly vinyl acetate and polybutenes are some examples for stickers used in herbicide formulations The spreader part is a surfactant that break the surface tension on the leaves and thus increase the spray coverage. When they are combined the product is called as spreadersticker and it is used in foliar applied herbicide formulations. By the addition of spreader sticker, effectiveness of herbicide application is increased and thus decreasing the chemical usage.

Drift Inhibitors/Thickening Agents

Drift control agents are used to reduce spray drift, which usually occurs when fine (< 150 μm diameter) spray droplets are carried away from the target area by wind and contaminate adjacent areas. In aerial spraying, spray nozzles create small droplets that travel faster than larger droplets. In such cases drift retardants are used to increase the mean droplet size. These adjuvants help to reduce off target drift of herbicides. Drift control agents alter the viscoelastic properties of the spray solution, yielding a coarser spray with greater mean droplet sizes and weights, and minimizing the number of small, easily-wind borne droplets (Hewitt, 1998). These agents are typically composed of large polymers such as polyacrylamides, and polysaccharides, and certain types of gums.

Compatibility Agents

Compatibility agents prevent chemical and/or physical interactions between different herbicides and fertilizers when these compounds are combined. These are adjuvants that allow easier mixing of two or more components in a solution. For instance, if the herbicides bentazon and sethoxydim are mixed, they may react to form precipitates, resulting in reduced rates of sethoxydim penetration (Wanamarta et al. 1993). In most cases, the herbicide label will state which herbicides may or may not be mixed together. When 2,4-D is applied with liquid-nitrogen fertilizers the solution may separate even if mixed vigorously unless a compatability agent is added to the mixture.

pH Buffers

pH plays a large role in herbicide efficacy pH of tank mix affects the half-life, solubility and efficacy of the herbicide, and may determine whether or not precipitates are formed (McMullan, 2000). Buffering agents help to raise or lower pH of spray mixtures to the designed pH of the formulation. Quantity required in a spray tank is difficult to determine. It depends on the pH of water used and also the acid value of the formulation.

Acidifiers

These are chemicals that can reduce the pH of solutions. They are useful in places where the pH of water is high. However, indiscriminate use of these chemicals is not recommended. Optimum pH is required for effectiveness of the chemical.

De-foamers

Excessive foaming can be a problem with agitation systems in sprayers, especially when the water level in the tank is low. Defoamers are adjuvants that reduce foaming in the spray tank. These are usually silicone containing products that are used in relatively small amounts.

Herbicide Safeners or Antidotes

Herbicide safeners are chemical compounds that selectively protect crop plants from herbicide damage without affecting the activity on target weed . Benefits offered by the use of safeners include (1) selective control of weeds in botanically related crops (2) possibility of selective control of weeds by nonselective herbicides (3) nullification of residual activity of persistent herbicides like triazines in soil etc. The concept to enhance crop tolerance to nonselective herbicide by using chemicals was established by Otto Hoffman in the late 1940s. Hoffmann observed no herbicide injury symptoms in tomato plants previously treated with 2,4,5-T, but when the plants were exposed accidentally to vapors of 2,4-D due to the malfunction of the ventillation system of the greenhouse, toxicity symptoms were observed. The idea of using safeners in the herbicide formulation practically started with the introduction of 1,8- naphthalic anhydride (NA) to improve the tolerance of maize to thiocarbamate herbicides.

Many safeners of different classes viz., naphthopyranone derivatives, chloroacetamides, oxime ether, thiazole carboxylates etc. are commercially used in combination with herbicides. The safener fenclorim of phenylpyrimidine class is used along with pretilachlor so as to protect rice crop. Naphthalic anhydride (NA) or dichlormid protected maize , sorghum and rice from the injury caused by chlorosulfuron. Use of NA also protected the crops from damage by imidazolinones, (e.g. Imazaquin). Several safeners are formulated as seed dressings eg. NA, cyometrinil etc. The effectiveness of safeners are more visualised when they are mixed with herbicides and applied to soil.

Safeners alleviate the crop damage caused by herbicides through several mechanisms viz., enhancing the metabolism of herbicides in crops, affecting absorption and transportation of herbicides in crops, competitively binding to herbicide target sites, and affecting activity of target enzymes. However, the exact mechanisms of action of safeners are still unclear. However, the action of NA in counteracting the phytotoxicity of chloroacetanilide and thiocarbamate herbicides to maize demonstrated that the safener markedly increased the levels of enzymes namely glutathione -S – tranferase and glutathione which are involved in detoxification of herbicides.

The list of commercially available herbicide safeners is given in Table

Table 1: Commercially available herbicide safeners

Safener		Herbicides counteracted	Crops protected	Application method
Chemical class	**Name**			
Anhydride	1,8- naphthalic anhydride (NA)	Thiocarbamtes (EPTC, Butylate)	Maize	Seed treatment
Dichloroacetamide	Dichormid	Thiocarbamates, Chloroacetanilide	Maize	Pre-plant incorporated with herbicide
	AD-67	Acetochlor	Maize	Spray (Pre –emergence)as a mixture with herbicide
	Benoxacor	Metolachlor	Maize	Spray (Pre –emergence)as a mixture with herbicide
Oxime ether	Cyometrinil	Chloroaetanilide (Metolachlor)	Sorghum	Seed treatment
	Oxabetrinil	Chloroaetanilide (Metolachlor)	Sorghum	Seed treatment
		Chloroaetanilide (Metolachlor)	Sorghum	Seed treatment
Thiazole carboxylic acid Dichloromethyl ketal	Flurazole	Alachlor	Sorghum	Seed treatment
	MG-191	Thiocarbamates, chloroacetanilide	Maize	Spray (Pre –emergence)as a mixture with herbicide

Safener		Herbicides counteracted	Crops protected	Application method
Chemical class	Name			
Phenylpyrimidine	Fenclorim	Pretilachlor	Rice	Spray (Pre –emergence)as a mixture with herbicide
Urea	Dymron	Pretilachlor, Pyrazosulfuron ethyl	Rice	Spray (Pre –emergence)as a mixture with herbicide
Piperidine-1-carbothioate	Dimepiperate	Sulfonylureas	Rice	Spray (Post –emergence)as a mixture with herbicide
8-Quinolinoxy carboxylic esters	Cloquintocet-mexyl	Clodinafop-propargyl	Cereals	Spray (Post –emergence)as a mixture with herbicide
1,2,4-Triazole carboxylate	Fenchlorazole-ethyl	Fenoxaprop-ethyl	Cereals	Spray (Post –emergence)as a mixture with herbicide
Dihydropyrazole-dicarboxylate	Mefenpyr-diethyl	ACCase inhibitors (Sulfonyl ureas)	Wheat, Rye, Barley	Spray (Post –emergence)as a mixture with herbicide
Dihydroisoxazole -carboxylate	Isoxadifen –ethyl	ACCase inhibitors (Sulfonyl ureas)	Maize, Rice	Spray (Post –emergence)as a mixture with herbicide
Arylsulfonyl-benzamide	Cyprosulfamide	Isoxaflutole	Maize	Spray (Pre –emergence)as a mixture with herbicide

Sources: Stephenson and Yaaocoby(1991), Davies(2001), Rosinger(2014)

References

Agricultural chemical label manual. Rohm and Haas Company, Independence mail, West Philadelphia,PA.

Davies, J. 2001.Herbicide safeners-commercial products and tools for agrochemical research. Pesticide outlook. February:11-15

Hewitt, A.J., Solomon, R.K., Marshall,E.J., 2009. Spray droplet size, drift potential, and risks to nontarget organisms from aerially applied glyphosate for coca control. Colombia. Journal of Toxicology and Environmental Health Part A, v.72, n.15-16, p.921-929

Hoffmann O. L. 1978. Herbicide antidotes: From concept to practice. In: Chemistry and action of herbicide antidotes. Pallos F. M., Casida J. E. (Eds). pp 35-61, Academic Press, New York, NY, USA

http://www.ncbi.nlm.nih.gov/pubmed/19672761>. Acesso em:

https://www.semanticscholar.org/author/G.-Wanamarta/90266195

Lori, T.H. 1992. A guide to agricultural spray, surfactants used in the United States. Thomas Publication, P.O. Box 9335. Fresno, CA. 244p.

McMullan, PM (2000) Utility adjuvants. Weed Technol 14:792–797

Rosinger, C. 2014. Herbicide safeners: an overview.26th German conference on weed biology and weed control. March11-13, 2014, Braunschweig, Germany

Stephenson , G. R. and Yacooby, T. 1991. Milestones in the development of herbicide safeners. Z.Naturforsch. 46c.:794-797

Wanamarta, G., Kells, J.J., Penner, D.1993. Overcoming Antagonistic Effects of Na-Bentazon on Sethoxydim Absorption.Weed technology.Vol 7.322-325

9

Herbicide Absorption and Translocation

Herbicides have received wide acceptance from farmers mainly due to ease of application and effectiveness. After application, herbicide molecules should penetrate plant tissue and get translocated to the target site. Once they reach the target protein they bind to them and disturb metabolic processes in plants by disrupting biosynthetic pathways and damaging cell structures by generating cytotoxic molecules that ultimately kill the plants. In this, the first step is penetration of herbicide into the plant system. Absorption of herbicide depends on application technique, phytochemical properties of the formulation such as lipophilicity and acidity of herbicides and also characteristic feature of plants which comes in contact with the chemical. Herbicides enter the plants either through leaf or root or through both.

Mechanism of action refers to the primary biochemical or biophysical reactions which bring about the ultimate herbicidal effects.

Mode of action refers to the sequence of events from absorption into plants to plant death.

Soil Applied Herbicides

Herbicides are applied to the soil as pre plant incorporation or pre-emergence application so that the germinating weed seeds or young germinating seedlings will take up the chemical either through root or shoot. Here the absorbing surface will either be the root hairs or the cortex cells. Most of the herbicides are absorbed by the area a few millimeter behind the root tip. Herbicide entry is by mass flow or by diffusion against a concentration gradient. Soil applied herbicides can be divided into four categories based on site of absorption.

1. **Herbicides that are absorbed by roots only and have activity at root meristems.** These herbicides are not translocated to shoot as they are lipophilic and so they are partitioned into cell membrane and nearby lipophilic areas. Water solubility is very meagre so their movement in soil is very limited. Pendimethalin belongs to this category of herbicide.

2. **Herbicides that are absorbed by both roots and shoots with some movement in developing seedling:** They are water soluble herbicides absorbed by roots and below-ground shoot tissue and have some ability to translocate within the seedling before emergence. Metolachlor is one herbicide which belongs to this category. It is absorbed by germinating seed along with water by mass flow and kills the emerging seedlings.

3. **Herbicides that are absorbed primarily by the shoot in the vapor phase and have some movement into developing seedlings.** Herbicides with low water solubility but high vapour pressure tend to accumulate as gas or vapour in the soil free space. As plant shoots are lipophilic they are absorbed by shoots in the gaseous form, Eptam and other herbicides in the carbamothioate group belong to this category.

4. **Herbicides that are absorbed by roots, but do not have herbicidal activity in the root system.** In case of herbicides like Atrazine, they are absorbed by root but the target site is the shoot and so they should be translocated from root to shoot.

Foliage Applied Herbicides

Foliar applied herbicides are intercepted by leaves and moves through phloem. Major barriers of foliar applied herbicides is the cuticle which is a complex matrix of materials that vary in water solubility and includes waxes, cutin and pectin. It consists of three layers which are epicuticular waxes, cuticle proper, and cuticular layer. These different layers are made up of polysaccharides, cellulose, cutin, and waxes. The lipophilic (fat loving) layer of the cuticle which includes epicuticular and cuticular waxes is the major transport limiting fraction of the leaves as they prevent penetration of foliar applied hydrophobic herbicides. Initial absorption into cuticular wax will depend on the percentage of total herbicide intercepted by the leaf. Lipophilic herbicides such as atrazine, fluizafop-butyl, oxyfluorfen are readily absorbed into the cuticular waxes and move easily to cutin and pectin layers. Movement of lipophilic compounds becomes difficult once it reaches the epidermal cells which are hydrophilic. So most of the herbicides are formulated as esters to enhance absorption and later on gets cleaved and produce free acids which is the active form of the herbicide.

In case of hydrophilic compounds, movement into cuticular waxes and absorption is enhanced with surfactants and liquid fertilizers that help to dissolve surface waxes and slow the drying time of spray droplets Hydrophilic herbicides move readily through polysaccharides and cutin rather than embedded waxes. It becomes easier as these herbicides approach the pectin and cell wall layers. Hydrophilic herbicides have a difficult time crossing the lipophilic plasma membrane of the target cells. Some herbicides have functional groups mainly carboxylic acid which can change from a hydrophilic to lipophilic forms depending on the pH of the surrounding environment, this helps in **"ion trapping"**

Herbicide is also absorbed by stem of the plant. Presence of bark makes entry difficult for herbicides. So growth stage of the plant is an important criteria that decides the effectiveness of the herbicide.

Translocation of Herbicides in Plants

Herbicide translocation from root to shoot is primarily through xylem. Movement to xylem from soil through root hairs is through 'apoplast', 'symplast' or by 'transmembrane' transport. Movement through xylem is faster than phloem movement

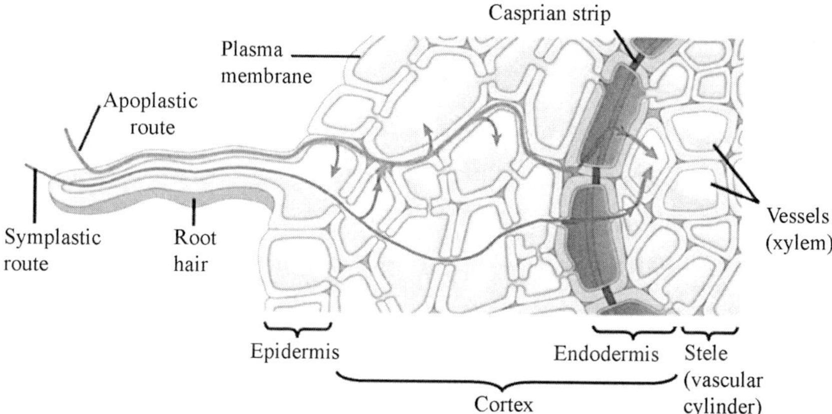

Fig. 1: Symplastic and Apoplastic Route for Herbicide entry through Root

Apoplastic Route

Movement of water, herbicides and minerals through the non-living portion (apoplast) of the plants is the apoplastic pathway. It includes groups of dead cells such as cell walls, intercellular spaces, and xylem - which form a continuum where water and solutes have the ability to move (Jachetta, 1986). This provides a route toward the vascular stele through free spaces and cell walls of the epidermis and cortex. In the endodermis there is the casparian strip which is a suberized layer that forces all herbicides to move in the symplast in order to enter the vascular system. An additional apoplastic route that allows direct access to xylem and phloem is along the margins of secondary roots.

Secondary roots develop from the pericycle which is a cell layer just inside the endodermis (Fig 1). Since secondary roots grow through the endodermis, a direct pathway to the xylem and phloem is available that bypasses the casparian strip and allows herbicides to enter the vascular system without moving into the symplast (living tissue). Movement through apoplast is faster.

Symplastic Route

According to Hay (1976) symplast is defined as the total mass of living cells in a plant. The symplastic (living) route forms a long and complex net along the plant both through phloem and through direct connections between neighboring cells by plasmodesmata. They are channels of cytoplasm lined by plasma membrane that transverse cell walls. These channels allow herbicides to move from cell to cell without passing through the cell wall. Plasmodesmata represent direct cytoplasmic connections from one cell to the next.

Mature leaves are the primary source of carbon assimilation and the assimilates move towards the sink through the symplastic pathway which is mainly through phloem. In foliar applied herbicides, the chemical moves from source to sink through the symplastic route. Sink is usually a region of rapid growth or a developing propagule. Here the movement can be either upward or downward depending on the demand for assimilates. Hence the rate direction and extend of phloem mobile herbicide is determined by the source sink relationship which varies among weeds depending on the growth stage and the life cycle.

Once the herbicide is translocated through xylem or phloem it spreads to the neighbouring cells in four ways.

1. Mass flow which is an apoplastic distribution.
2. Passive diffusion as per an electrochemical gradient.
3. Active translocation involving protein carriers at the expense of ATPs
4. Translocation through plasmodesmata

The trans-membrane route involves movement across cells and cell walls combining both symplastic and apoplastic movement. Many of the herbicides are classified as xylem or phloem mobile but in reality herbicide translocation is a complex process that may involve both xylem and phloem movement. This is true in case of atrazine which has equal access to both xylem and phloem. Chemical characteristics of atrazine allow rapid movement across plasma membrane giving equal access to xylem and phloem. The xylem and phloem cells are separated only by a single cell layer so it allows transmembrane movement ofherbicide. Movement ofherbicide through xylem is approximately twice the rate of phloem hence the net movement appears to be in the apoplast or xylem. So translocation of atrazine is also described as pseudo apoplastic movement.

Some soil-applied auxinic herbicides 2,4-D, MCPA and picloram are systemic and produce symptoms at the apical meristems of susceptible plants. These herbicides travel in the xylem to margins of older exporting leaves and then move into the phloem to translocate with sugars to the apical meristem

Factors Affecting Absorption and Translocation of Herbicides

Important factors that affect the absorption and translocation of herbicides are soil, plant, herbicide chemistry and environmental conditions..

Plant Factor

Plant factors that affect the absorption and translocation of herbicides are age of plant, leaf orientation and thickness of cuticle. Younger actively growing plants show more absorption and translocation of herbicides. In younger plants cuticle thickness is less and translocation to the growing meristematic region is faster. Presence of hairs and waxes on leaf surface prevents penetration of herbicides. Herbicide penetration is also less in narrow leaves.

Soil Factors

Pre plant and pre emergence herbicides applied to the soil get adsorbed to soil, dissolve in water or in some cases even get converted to vapour which indicates that it can be available in solid, liquid or gaseous phases. Concentration of herbicide in aqueous solution is an important determinant for absorption either by mass flow or diffusion. Availability of herbicide depends on soil texture, soil moisture, temperature and soil pH. Their influence on herbicides availability is given in detail in the portion on fate of herbicides in soil (Chapter 12).

Herbicide Factors

In soil applied herbicides their solubility and persistence in soil affects intake by plants. pH of the herbicide is an important factor that influences uptake by plants. Generally weak acids are more mobile in plant systems.

Most foliar herbicides are formulated with adjuvants which help in penetration of herbicide molecule into cuticular waxes. Properties of the formulations influence absorption and translocation process in plants.

References

Ashton, F.M. and Crafts, A.S. 1981. Mode of Action of Herbicides. 2nd Ed. Wiley-Inter Science.

Beevers, H., Goldschmidt, E. P. and Koffler, H. 1952. The use of esters of biologically weak acids in overcoming permeability difficulties. Arch. Bio-chem. 39: 236-238. 195

Briggs, G.G., Rigitano, R.L.O. and Bromilow, R.H. 1987.Physio-chemical factors affecting uptake by roots and translocation to shoots of weak acids in Barley. Pesticide Science 19:101-112

Carlson, W.C., Lignowski, E.M. and Hopen, H.J.1975. Uptake, translocation and absorption of pronamide, Weed Sci. 23: 148-154

Das, T.K. 2009. Weed Science- Principles and Application. Jain Publishers.

Devine, M., Duke, S.O., Fedtke, C. 1993.Physiology of herbicide action. Englewood Cliffs: Prentice Hall, 441p.

Devine, M. D., Hall, L. M.1990. Implications of sucrose translocation mechanisms for the translocation of herbicides. Weed Sci., v. 38, p. 299-304.

Gutschick, V.P. 1999. Biotic and abiotic consequences of differences in leaf structure. New Phytol. 143: 3–18. doi:10.1046/j.1469-8137.1999.00423.x.

Hay, J. R. 1976. Herbicide translocation in plants. In: AUDUS, L. J. (Ed.). Herbicides physiology,biochemistry, ecology. V.1, p.365-393

Jachetta, J. J.; Appleby, A. P.; Boersma, L.1986. Apoplastic and symplastic pathways of atrazine and glyphosate translocation in shoots of seedling sunflower. Plant Physiol., v. 82, p. 1000-1007.

Katagi,T. and Mikami,N.2000.Primary metabolism of agrochemicals in plant.pp.43-106. In;Roberts,T. (ed.) Metabolism of Agrochemicals in Plants.John Wiley,New York.

Knoche, M. 1994. Effect of droplet size and carrier volume on performance of foliage-applied herbicides. Crop Prot. 13(3): 163–178. doi:10.1016/0261-2194(94)90075-2.

Mandal, R.C. 1990. Weed, Weedicides and Weed Control - Principles andPractices. Agro-Botanical Publ.

Rao VS. 2007. Principles of Weed Science. Oxford & IBH.

Ross MA & Carola Lembi A. 1999. Applied Weed Science. 2nd Ed. Prentice Hall.

Singh, R.K. and Das, T. K. 2016. Herbicide: Mode and mechanism of action and resistance. In Yaduraju, N. T. Sharma, A. R. and Das, T. K. (Eds.).2016. Weed Science and Management.402p. Indian Society of Weed Science, Jabalpur and Indian Society of Agronomy, New Delhi

Wang, C.J., and Liu, Z.Q. 2007. Foliar uptake of pesticides-present status and future challenge. Pest Biochem. Physiol. 87: 1–8. doi:10.1016/j.pestbp.2006.04.004.

10

Compatibility of Herbicides and Other Agrochemicals

In a cropping systems there may be simultaneous or sequential application of herbicides, insecticides, fungicides, fertilizers, PGR during a single cropping season. The different chemicals will undergo physical and chemical changes as they interact with plant, soil and other chemical substances already applied to the plant or soil. The interaction effect of such chemical substances will be visible only at the later end of the cropping season. The responses observed in the plants can be

i) Additive, when the total effect of a combination is equal to the sum of the effects of the components taken independently.

ii) Synergistic when the effect of a combination is greater or more prolonged than the sum of the effects of the two chemicals applied independently. Such effects have been observed when glyphosate was mixed with 2,4-D: effect on field bindweed (*Convolvulus arvensis*) was more as compared with separate applications (Flint and Barrett, 1989). In johnsongrass (*Sorghum halepense*) mixtures of sethoxydim with dimethenamid was more effective when compared with separate applications. (Scott et al. 2001)

iii) Antagonistic when the total effect of a combination is smaller than the effect of the most active component applied alone. In the case of crabgrass, (*Digitaria sanguinalis*) it has been observed that application of pyrithiobac in mixture with fluazifop-P reduces efficacy of fluazifop-P. Similarly when tribenuron is combined with diclofop it reduces efficacy of diclofop on wild oat (*Avena fatua*). Combining 2,4-D with cyhalofop butyl to broaden the effect to dicots and sedges was found to reduce the effect of cyhalofop. Since these combinations are less efficient they should be avoided.

iv) Independent effect, when the total effect of a combination is equal to the effect of the most active component applied alone.

v) Enhancement effect: The effect of a herbicide and non-toxic adjuvant applied in combination on a plant is said to have an enhancement effect if the response is greater than that obtained when the herbicide is used at the same rates without the adjuvant. Eg. mixing ammonium sulphate or urea (1-3%) with glyphosate solution enhances its efficiency. Similarly adding the adjuvant POEA in glyphosate formulation increases its efficiency.

Herbicide Combination and Interaction

Modern trend is to combine two herbicides to get better weed control. Herbicide mixtures have several advantages when compared with single herbicides. It increases the spectrum of weeds controlled and also extends the control for a longer period. The dosage used can be reduced and minimum dosage of the chemicals rather than high dose of a single chemical can give better control. This will also reduce the persistence of herbicides in soil and crop residues. Appearance of herbicide resistant weed species in crop fields can be delayed. It reduces soil compaction by avoiding multiple field operations and helps to save cost of cultivation. However, when two chemicals are combined, behavior of both the chemicals will be affected and so it is difficult to predict the efficacy of the chemical combination. Compatibility is the capacity for two systems/things to work/mix together without having to be altered to do so. A successful herbicide combination should have enhanced activity on the target weed and decreased activity on crop plants.

Interaction between two herbicides mainly depends on the chemical group, the absorption rate of the chemicals, physiological action of the herbicides and the metabolic pathway involved. This means that the herbicides can interact physically or chemically in the spray solution or biologically in the plant system.

Chemical compatibility is the most important factor. When the companion herbicides belong to the same chemical group synergism is often seen. Antagonism is more often observed when the herbicides belong to two different chemical groups. As they have different chemical structures, there are chances for them to interact and form inactive complexes. According to Zhang et al.(1995) there is three times more chances of antagonism than synergism in herbicide combinations.

Behaviour of herbicide mixtures depends on the point of entrance and the mobility of the combined herbicides into the plant. When two chemicals enter the plant simultaneously through the same point either root or foliage there are chances that one herbicide in the mixture may reduce the absorbed amount of the other and consequently reduce its efficacy. Similarly when they are translocated together to the same site of action one of the herbicide may reduce action of the other.

It has been observed that the efficacy of herbicide mixtures depends on the target weed species. When a combination of acifluorfen and bentazon was applied to Lambsquarters (*Chenopodium album*) and Velvetleaf (*Abutilon theophrasti*) efficiency of weed control increased but when the same combination was applied to Jimsonweed (*Datura stramonium*) and Red root pigweed (*Amaranthus retroflexus*), efficacy of the chemical was much lower

Stage of the weed is another important factor that affects the interaction between combined herbicides. Efficacy of chlorsulfuron and diclofop depended on the growth stage of Italian ryegrass (*Lolium multiflorum*). According to Liebl and Worsham (1987) the effect was more severe at two-leaf growth stage than at three-leaf growth stage. The reason may be lower detoxification ability at younger stage of the weed. Thinner cuticle at two leaf stage may probably allow greater absorption of the herbicide thereby improving the retention, absorption, and translocation of the chemicals.

A. Successful Combination Herbicide Formulations in the Market

Herbicide combination	Commercial Names
Chlorimuron ethyl+Metsulfuron methyl	Almix
Pretilachlor + Bensulfuron	Londax power
Cyhalofop butyl+Penoxsulam	Vivaya
Triafamone+Ethoxysulfuron	Council activ
Butachlor+Penoxsulam	Coreon
Florpyroxifen benzyl+Cyhalofop butyl	Novelet
Imazethapyr+Imazamox	Odyssey
Pretilachlor+Pyrazosulfuron ethyl	Swatch, Eros

B. Tank Mix Combinations

2,4-D + Glyphosate

2,4-D + Paraquat

Diuron + Paraquat

Oxyfluorfen + Glyphosate

Tank Mix Combinations are successful combinations of two herbicides which can be combined and used by the farmers. These are tried and tested synergistic combinations.

Herbicide –Insecticide Interaction

Combining herbicide with insecticide or fungicide is preferred by farmers as it saves the cost of dual application. In many cases the combination is antagonistic and causes crop injury. This is mainly because some of the pesticide chemicals may render the crop unable to metabolize and detoxify the herbicide. In some cases even sequential application of the two chemicals may show toxicity symptoms.

Application of some organophosphates such as counter or thimet for control of corn rootworm in combination with ALS inhibitor herbicides such as Accent (Nicosulfuron) or Beacon (Primisulfuron) can injure corn significantly. Symptoms of this injury can be observed as stunting, yellowing, chlorosis and bleached bands on leaves.

It was observed that pyrethroid insecticides can be substituted for organophosphates as they do not affect efficacy of post emergence herbicides. When acephate was mixed with sethoxydim no adverse effect was noted. Similarly application of chlorpyrifos at the time of planting did not affect peanut response to pre-emergence application of diclosulam, S-metolachlor, or flumioxazin and even to post-emergence application of acifluorfen or acifluorfen plus bentazon.

In case of fungicide Dexon, it was observed that in corn, cucumber and soybean there was antagonistic interaction with atrazine as dexon reduced uptake of atrazine.

Many studies have reported that herbicide application can have a positive or negative impact on diseases of field crop. According to Sharma and Sohi (1983) application of bromacil, diuron, nitrofen, and alachlor all these herbicides reduced disease severity in *Phaseolus vulgaris* by *Rhizoctonia.* Studies by Caulder et al., (1987) showed that use of herbicides such as bentazon, acifluorfen, chlorimuron, fluazifop, diclofop, sethoxydim, imazaquin, metribuzin, oryzalin, thidiazuron, diaminozide, and mefluidide enhanced disease severity of plant pathogens such as *Alternaria cassiae, Colletotrichum coccodes, C. truncatum, and Fusarium lateritium*

This may be because some herbicides such as Glufosinate have been reported to have fungitoxic and antimicrobial activity. It was also reported that pretilachlor and butachlor trigger accumulation of the phytoalexins momilactone A and sakurantetin in rice leaves while pendimethalin, induces the synthesis of the phytoalexin tomatine in tomato.

Herbicide- Fertilizer Interactions

In developed countries, application of fertilizers along with herbicides is becoming popular. In fertigation process combining herbicides with fertiliser suspension does not produce any detrimental effect on plants. In case of atrazine it was noted that when combined with NPK it showed minimum phytotoxicity symptoms on crop plants. With P and K alone there was more phytotoxicity symptoms but when P or K was applied with N, phytotoxicity symptoms was lower.

In case of 2,4-D and glyphosate it was seen that combining with urea or ammonium sulphate increased efficacy. Herbicide dicamba, decreased the uptake of nitrogen. It has been noted that benefit from addition of ammonium is observed in relatively polar, weak acid herbicides such as basagran, sulfonylureas and imidazolinones. Addition of Ammonium-based fertilizers and, in particular, ammonium sulfate (AMS) is found to reduce potential antagonism with hard water or antagonism with other pesticides. In case of roundup in the label it is given that addition of ammonium sulphate will improve efficacy of the chemical when hard water or drought condition is noticed.

When micronutrients such as boron and manganese are applied it is seen that occasionally it affects the performance of the herbicides. This is especially true in crops like peanut. It was observed that efficacy of the herbicide clethodim and imazethapyr was reduced by micronutrient application when some weeds were evaluated.

Growth regulator herbicides belong to three categories viz. phenoxy (2,4-D; 2,4-DP; MCPA; MCPP), benzoic acid (dicamba), or pyridine (triclopyr). In extremely small doses these synthetic auxin herbicides stimulate plant growth, but in proper concentrations, synthetic auxins disrupt numerous biochemical pathways and will kill susceptible plants. They are selective in their action and kill only dicot weeds even at high concentrations. Efficacy of the herbicides like acifluorfen, acifluorfen plus bentazon, bentazon, imazethapyr, imazapic, lactofen, and 2,4-DB was not affected by prohexadione calcium which is used as a growth regulator in peanut (Beam et al., 2002).

References

Beam, J.B., D.L. Jordan, A.C. York, J.E. Bailey, T.G. Isleib, and T.E. McKemie. 2002. Interaction of prohexadione calcium with agrichemicals applied to peanut (*Arachis hypogaea* L). Peanut Sci. 29:29-35

Brandenburg, R.L., Walls, F. R., Casteel, S. and Hudak, C. 2006. Influence of application variables on efficacy of boron-containing fertilizers applied to peanut (*Arachis hypogaea* L.). Peanut Sci. 33:104-111.

Caulder, J.D., Gotleib A.R., Stowell, L., and Watson, A. K. 1987. Herbicidal compositions comprising microbial herbicides and chemical herbicides or plant growth regulators. Eur. Pat. Appl. CODEN: EPXXDW EP 207653 A1 19870107 CAN 106:80406 AN 1987:80406 39 pp.

David, J. Chahal, G.S., Lancaster, S. R. Beam, J., York, C. A. and Reynolds, W.N. 2011. Defining Interactions of Herbicides with Other Crop Protection Products Applied to Peanut. In book: Herbicides, Theory and Applications: 73-92

Diggle, A. J.; Neve, P. B.; Smith, F. P. 2003. Herbicides used in combination can reduce the probability of herbicide resistance in finite weed populations. Weed Res., V. 43, n. 5, p. 371-382.

Duke, S.O., D.E. Wedge, A. L. Cerdeira, and M. B. Matallo. 2007. Herbicide Effects on Plant Disease. Pest Manage. Sci. 18:36-40.

Flint, J. L. and Barrett, M. 1989. Interactions of glyphosate with 2,4-D or dicamba on field bindweed (*Convolvulus arvensis*). Weed Sci. 37:12–18.Google Scholar

Flint, J. L. and M. B. Barrett. 1989b. Antagonism of glyphosate toxicity to johnsongrass (Sorghum halepense) by 2,4-D and dicamba. Weed Sci. 37:700-705.

Green, M. J. 1989. Herbicide antagonism at the whole plant level. Weed Technol. 3:217-226.

Grichar, W. J. 1991. Sethoxydim and broadleaf herbicide interaction effects on annual grass control in peanuts (*Arachis hypogaea*). Weed Technol. 5:321-324

Hatzios, K. K. and D. Penner. 1985. Interaction of herbicides with other agricultural chemicals in higher plants. Rev. Weed Sci. 1:1-64.

Jordan, D. L., S. H. Lancaster, J. E. Lanier, P. D. Johnson, J. B. Beam, A. C. York, R. L. 2013. Influence of Application Variables on Efficacy of Boron-Containing Fertilizers Applied to Peanut (*Arachis hypogaea* L.) Peanut Science .V.33(2).104-111

Liebl, R. and Worsham, A.D. 1987. Interference of Italian Ryegrass (*Lolium multiflorum*) in Wheat (*Triticum aestivum*). Weed Science, 35, 819-823.

Scott, G. H., Askew, S. D., and Wilcut, J. W. 2001. Economic evaluation of diclosulam and Flumioxazin systems in peanut (*Arachis hypogaea*). Weed Technol. 15:360-364.

Sharma S and Sohi H. S.1983. Influence of herbicides on root rot of French beans (Phaseolus vulgaris L.) caused by Rhizoctonia solani Zentralblatt fur Mikrobiologie Vol1355:357-61

Zhang, J. H., Hamill, A. S., Weaver, S. E.1995. Antagonism and synergism between herbicides-trends from previous studies. Weed Technol., V. 9, n. 1, p. 86-90.

11

Formulations Dosage and Application of Herbicides

Major advantage of chemical herbicides over other methods of weed control is its ease of application, which saves the cost of labour for the farmer. Effectiveness of herbicide depends on how long it remains in active form in the plant /soil. The form in which any herbicide is available in the market is known as herbicide formulation. The technical active ingredient (ai.) of the herbicide is converted to a formulations so as to improve easiness of handling, retention on leaf surface, absorption and movement in the plant. Herbicides are available in different formulations. When the technical active ingredient (molecule responsible for herbicide effect) is made ready to be used by mixing with liquid or dry dilutents either by grinding or by addition of emulsifiers, stabilizers or other adjuvants like activators, safeners etc., it becomes a formulation. Development of a formulation depends on the physiochemical properties of the active ingredient and inert materials used in the process. The formulated product should have chemical stability and physical properties like suspendability for wettable powders and no flocculation, sedimentation or creaming. Emulsifiable concentrates should have stability so that there will be no separation of formulation components in the colloidal system. Stable particle size is important to maximize biological activity, flowability, adhesivity, deposition on plant surface, suspendability, density and desired partitioning co-efficient etc.

Types of Formulation

Generally herbicide formulation available in the market can be in the form of Wettable powder (WP), Soluble Powder (SP), Soluble Liquid (SL), Emulsifiable Concentrate (EC), Water Dispersible Granule (WDG) or Water Soluble Powder (WSP) Flowable Suspension (FS) Colloidal Suspension (CS) Dusting Powder (DP), Dispersible Solid (DS), Granule (G/GR), Suspension Concentrate (SC), Soluble Granule (SG), Soluble Liquid (SL), Wettable Granule (WG), Water Soluble Tablets (WS), Fumigants and Nano herbicides.

Wettable Powder (WP)

Herbicide formulation is made into a Wettable powder (WP) when the solubility of the active ingredient is limited and it is impossible to make a solution or an emulsifiable concentrate economically. It contains finely divided solids (<3 μm) that can be readily suspended in water to form a suspension. In this case the amount of active ingredient (ai.) loaded into the formulation can be low (<30%) medium (30-60%) or high (60-90%).

This type of formulation contains active ingredient (ai), dilutents, surface active agents and an axillary material. Low levels of ai. in such formulations, prevents caking. Dilutents will have low bulk density and relatively large sorptive capacity. Commonly used dilutents are silica gel, hydrated Al_2O_3, synthetic calcium silicate, talc, kaolinite, attapulgite and bentonite diatomaceous earth. Detergents act as surface active agents. Sulphonates of alkali metals, Sodium diphenyl oxide disulfonate, sodium dodecyl benzene sulphonate, alkyl aryl ethers of polyethelene glycol and polypropylene glycol and sodium alkane sulphonate are widely used as surfactants. Auxillary materials are sodium salts of lignin sulpho acids and sodium salts of sulpho acids obtained by sulphonation of petroleum products. The wetting agents and stickers used can be carboxy methyl cellulose, methyl cellulose, polymers of unsaturated alcohol, gelatin, animal glues, caseinates, salts of resin acids etc. WP formulations have to be protected from contact with moisture.

Soluble Powder

These formulations are very similar to wettable powder, except that the ingredients used completely dissolve in water or the liquid for which the herbicide has been formulated. When mixed with water or the recommended liquid it becomes a true solution.

Emulsifiable Concentrate (EC)

Emulsification is normally done to disperse small volumes of the concentrate in large volumes of the spray carrier. When they are diluted with water they give a stable emulsion suitable for spraying. Emulsions with equal concentration of active ingredient are more effective than the corresponding suspensions. Generally, two types of EC are made

Concentrated Emulsion

This is prepared by dispersion of tiny microscopic droplets of the ai of herbicide in an aqueous solution such as water. It is a uniform mixture of two immiscible solvent by agitation, whisking or blending to form a stable emulsion. Only herbicides which are stable to the action of water are suitable. Homogenized concentrated emulsions withstand low temperatures and prolonged storage. They may sometimes thicken in storage, but a stirring will be enough to bring them to normal condition.

Miscible Oil

They contain herbicides, solvent and emulsifier. Hydrocarbons and their halogenated derivatives, esters, various petroleum products, creolin, coaltar, oils may be used as solvent. Here the emulsifier acts as the surface active agent which will readily emulsify the concentrate in the final spray liquid. At times two or more emulsifiers are used.

Solution Concentrate (SC)/Water Soluble Concentrate (WSC)/ Soluble Liquids (SL)

In these formulations the herbicide is dissolved in the solvent system to provide a concentrate that is soluble in the carrier.

Granules (GR) The active ingredient (ai.) is coated on an inert material of appropriate particle size. To prevent carrier from getting dispersed in water the formulation is recalcinated to temperatures of 300°C and higher. Granules do not have drift problems and contribute to low foliage injury. Usually concentration of active ingredient in a formulation ranges from 1 to 20%. Mobility of herbicide from the granule is an important aspect that decides bioactivity of the chemical.

Fumigants

These are volatile chemicals applied to confined space or into the soil to produce a gas that will destroy weed seeds and act as a soil sterilants.

Nano Herbicides

In these formulations the ai. is coated with various materials of different sizes in nano range (1nm to100nm), so that they are released in a controlled way to get long term weed control.

Appliances and Application Techniques

Though most herbicides are considered nontoxic to animals and humans, when applied aerially, they can cause substantial mortality of non-target plants and insects. Equipments like sprayers, soil incorporation equipment, and spreaders for pelleted herbicides are available for improving efficiency and convenience of herbicide application.

Haudraulic (pressurised) applicators need water to work. Lever operated knapsack sprayers, motorised or ground driven applicators and wick applicators for contact application are generally used for spraying herbicides.

Nozzles

Nozzle selection is an important decision for effective herbicide application. Nozzle type affects the amount of spray applied to a particular area, uniformity of the applied spray, coverage on the sprayed surfaces and drift of the spray solution. Different nozzle types have specific characteristics and capabilities and are designed for use under different conditions. Nozzle types commonly used for ground application of agricultural chemicals are flat-fan, floodjet, and cone nozzle. Primary improvement in the development of nozzles has been to minimize the formation of driftable droplets (< 200 micron) while still producing a droplet spectrum that provides the level of coverage required for consistent weed management. Regular flat fan nozzles are used for spraying most of the herbicides. Floodjet nozzle create wide angle flat spray pattern and are clog resistant. They are ideal for high application rates and speed. Both flat fan and floodjet nozzles are available in different spray angles Viz.65, 80 and 110 degrees. These nozzles produce a flat oval spray pattern with tapered edge. They can also be fixed on a boom and used in mechanical spraying using a tractor. In case of systemic and translocated herbicides, thorough wetting of stem and leaves is required. So an impact or fan nozzle is recommended. In case of contact herbicide spray, a cone nozzle or fan nozzle or impact nozzle operated at 275 kpa will be more suitable.

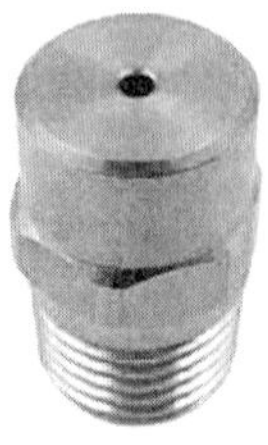

Flat fan nozzle   Flood jet Nozzle   Cone Nozzle

Mixing

Herbicide efficiency depends on thorough mixing of the active ingredient with the carrier (water). Steps involved in mixing

1. Fill the spray tank about half with water add the required quantity of the herbicide and agitate the solution to prevent the herbicide formulation settling.

2. Then fill the tank completely with water.
3. Mix the contents of the knapsack sprayer tank by shaking.

Operation of Knapsack Sprayer

A spray is defined as a liquid discharged into particles and scattered as dispersed droplets. Herbicide application is directed to soil in case of pre-planting and pre-emergence herbicides and to foliage in case of post-emergence herbicides. Compressed air (70-275 Kpa) forces the liquid from the pressure chamber through the hose to the nozzle. The nozzle turns the liquid into droplets, which aid in uniform coverage of weed or soil.

Spraying

Spraying should not be done when the wind is too strong. If it is essential, hold the nozzle close to the ground (within 50cm) to prevent droplets from being blown away during calibration of sprayer and during actual field spraying. Walking downwards is preferred as it will prevent the spray blown off from the crop reaching the person who is spraying the chemical.

1. Use soft materials to clean blocked nozzles. Never clean nozzle with a wire or pin.
2. A nozzle height of 50cm from the target is ideal. Avoid missing strips or overlapping spray swaths.
3. While spraying maintain a constant walking speed.

Cleaning Sprayer

1. Rinse the sprayer with clean water to remove most of the chemical. A household detergent can be added to water used for cleaning. Pour this out and add clean water again.
2. Operate the pump for a minimum of 10 strokes and pour out the remaining water.
3. Repeat the procedure at least twice. Presence of herbicide residue in the sprayer may be harmful to the crop when the sprayer is used for other purposes.

Herbicide Drift

Herbicide drift is a problem for sensitive crops like tomato, cotton, lettuce and even fruit trees. Symptoms of herbicide drift will be easily mistaken for disease or pest attack. Drift occurs when smaller droplets in the spray are carried away from the target by wind or vapour from a volatile herbicide in the atmosphere during or after spraying and reaches the nearby crop. Highest growth damage is observed in case of growth regulator herbicides like 2,4-D, MCPA or even glyphosate. This can be avoided by taking the following precautions.

1. Spray only when a slight wind is blowing away from a susceptible crop.
2. Use large nozzle tips, so that the droplet size is larger and also more spray volume (more than 10 gallons or 38 liters/ha).
3. Use minimum pressure to operate nozzle
4. Hold spray nozzle close to target.
5. Leave a 10m wide strip unsprayed between the sprayed crop and the field where a sensitive crop is grown.

Application of Granular Herbicide

These formulations are broadcasted in the field. However to ensure safety and uniformity of application the following precautions are recommended.

1. Select suitable dilutents for granules such as sand, dried farmyard manure, soil etc. which does not react with the herbicide and imparts physical stability to the granules.
2. Mark 25m^2 area, apply using a perforated can or manually, uniformly over the measured area.
3. Assess quantity of dilutents required. Work out the quantity required for the whole field.
4. Mix the herbicide thoroughly with dilutents for getting a uniform spread.

Water for Spraying

To work out the volume of water required to cover a measured area first conduct a dummy trial, since the speed of walking varies from person to person.

Step 1. Spray on a measured area (25m^2) at constant working speed at recommended pressure of pump.

Step 2. Asses the quantity of water sprayed and workout for the entire field.

Step 3. While adding water to the tank, keep the strainer to remove dirt or lumps in the water.

Step 4. Cover the field by spraying at recommended pressure and at constant speed to get uniform coverage.

Step 5. Clean the nozzle with soft material like matchstick if blocked in the middle of the spraying operation.

Dosage Calculation

Selection of herbicide and dosage calculation is important to get effective weed control. Informations required for dosage estimation are the following

a. Recommended dosage of herbicide for the crop.

b. Amount of active ingredient (ai.) in a given quantity of the commercial product. ai is defined as that part of a chemical formulation that is directly responsible for herbicidal effect.

$$\text{Quantity of formulated product per hectare} = \frac{\text{Recommended amount of ai}}{\text{Percentage ai in formulated product}} \times 100$$

E.g. To apply 2.0kg ai of Machete which has 60% ai as butachlor

The calculation is

$$\frac{2.0}{60} \times 100 = 3.333\text{kg of Machete/ha.}$$

In many cases the area to be sprayed may not be exactly 1 ha and the herbicide tank may not be big enough to hold all the herbicide and water required for 1 ha. The amount of water required to spray a given area can be estimated using the formula given below.

$$\text{Amount of water required (litres)} = \frac{\text{Area to be sprayed } (m^2) \times \text{sprayer output in litres/ha.}}{10{,}000}$$

To spray $1000m^2$ with a sprayer output of 200l/ha

Water required will be

$$\frac{1000 \times 200}{10{,}000} = 20 \text{ litres.}$$

Effectiveness of herbicide used in controlling weeds is estimated by using the parameters, Weed control efficiency (WCE) and Weed index. These parameters provides support in impact assessment, interpretations and drawing appropriate conclusions in weed management research.

Weed Control Efficiency (%)

Weed control efficiency (WCE) denotes the magnitude of weed reduction due to weed control treatment. It is worked out by using the formula suggested by Mani et al. (1973) and expressed in percentage.

$$\text{WCE (\%)} = \frac{\text{Dry wt. of weeds in un weeded plot- Dry wt. of weed in treatment plots}}{\text{Dry wt. of weed in un weeded plot}} \times 100$$

Weed Index

Weed index is the measure of the efficiency of a particular treatment when compared with the weed free treatment. It is defined as the magnitude of yield reduction due to presence of weeds in comparison with weed free check. In other words, weed index expresses the competition offered by weeds measured by per cent reduction in yield owing to their presence in the field (Gill and Vijayakumar, 1969). Higher weed index mean greater loss. Calculated by using following formulae

$$\text{Weed index (\%)} = \frac{x-y}{x} \times 100$$

Where,

x = Yield from the weed free check

y = Yield from the treatment for which weed index has to be calculated

Weed control efficiency indicates the bioefficacy of a treatment to control weeds. It does not account for the effect of the treatment on crop growth and yield. Sometimes herbicides used may cause some phytotoxicity or damage to the crop resulting in reduction in crop yield. This also is accounted in weed index.

References

Bhan,V.M., Singh, V.P. and Dixit Anil.1996. Weed Management-An important tool in improving crop production.NRCWS.Jabalpur.pp.228

Chattopadhyay, S.B.1980. Principles and procedures of plant protection. Oxford and IBH Publishing Co. New Delhi,pp.480

Gill, G.S. and Vijayakumar 1969, "Weed index" A new method for reporting weed control trials. Indian J. Agron., 16, 96-98

Herbicides Market report -Herbicides Market - Growth, Trends, and Forecasts (2023 - 2028) https://www.mordorintelligence.com/industry-reports/global-herbicides-market-industry

Kwesi Ampong Nyarko and De Datta, S.K.1991.A hand book for weed control in rice. IRRI, Manila, pp.113.

Mani, V.S., Malla, M.L., Gautam, K.C. and Bhagwndas. 1973. Weed killing chemicals in potato cultivation. Indian Farm., VXXII, 17-18.

Milberg, P. and Hallgren, E. 2004. Yield loss due to weeds in cereals and its large scale variability in Sweden. Field Crop Research 86: 199 – 209.

Paul, E. Sumner 2009. Sprayer Nozzle Selection. Cooperative extension publication. University of Georgia. https://secure.caes.uga.edu/extension/publications/files/pdf/B%20 1158_3.PDF

Rana, S.S. and Rana, M.C. 2015. Advances in Weed Management. Department of Agronomy, College of Agriculture, CSK Himachal Pradesh Krishi Vishvavidyalaya, Palampur, 183 pages (DOI: 10.13140/RG.2.2.26235.72487

Rao,V.S.1983. Principles of Weed Science. Oxford &IBH Publishing Co.,New Delhi. pp: 540

Tu, M., Hurd, C. and J.M. Randall. 2001. Weed Control Methods Handbook, The NatureConservancy, http://tncweeds.ucdavis.edu, version: April 2001.

12

Fate of Herbicides Applied in Soil

When a herbicide is applied to the soil, it is subjected to various interactions with soil and surrounding environment which ultimately determines its persistence. Dissipation of these compounds from soil system is by three main processes viz., physical (volatilization, leaching and runoff), chemical (photochemical decomposition, adsorption by organic and inorganic colloids and chemical decomposition) and biological (microbial decomposition and plant uptake)

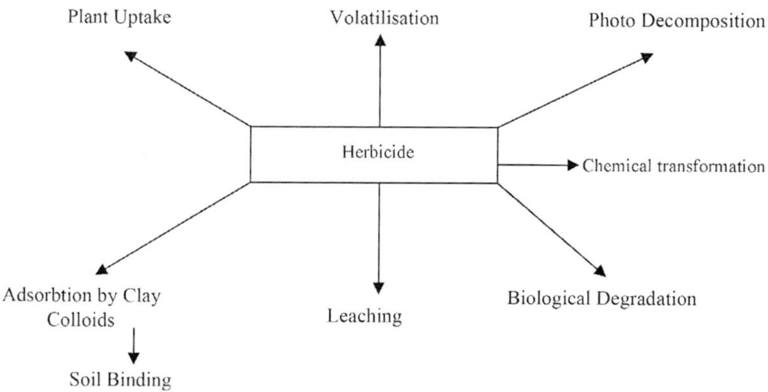

1. Physical Processes

a) Volatilisation

Chemicals which can easily change into vapour form move through the air into the atmosphere. Volatilisation of herbicide from soil or leaf surfaces depends mainly on vapour pressure of the compound, its concentration, adsorption to soil and solubility in water. It is also affected by air temperature, wind velocity above the soil surface, relative humidity, soil temperature, moisture as well as chemical nature of the compound. It is observed that volatilisation of pendimethalin is higher than that of terbuthylazine. Volatilisable herbicides are recommended for soil incorporation.

b. Leaching

Leaching of herbicides refers to their downward movement in soils as solutes in soil water. Leaching is more influenced by soil properties than by the physico-chemical properties of the compounds. As early as 1964, Hartley (1964) reported that percolating water appeared to be the principal means of movement of water-soluble compounds. The degree of leaching of a pesticide from soil is correlated with its water solubility, but this is also modified by the capacity of pesticides for adsorption on soil fraction (Edwards, 1973).

Degree of leaching of chemicals through soil decides ground water contamination. In paddy fields, Sodium salt of 2,4-D, pretilachlor and oxyfluorfen are some of the common herbicides. Among these, leachability of 2,4-D is much higher than the other two compounds. So residues of 2,4-D can enter ponds and streams by direct application, drift or run off from soils and reach the pond bottoms and irrigation channels especially if the rains occurred shortly after the application. Oxyfluorfen is strongly adsorbed on soil particles and leaching loss is very little. Leachability of pretilachlor is between 2,4-D and oxyfluorfen (Devi et al., 2015).

B. Chemical Transformations

1. Photochemical decomposition: Decomposition by light is one of the mechanisms of herbicide conversion in soil under field conditions and this process is induced by UV radiation. When exposed to light, organic molecules absorb radiant energy and cause excitation of electrons which can induce transformations such as breaking or formation of chemical bonds, fluorescence etc. The rate of decomposition and the nature of products depend on light source, intensity of light and the chemical structure of the herbicide. In case of diquat and paraquat, the molecules are characterised by weak bonds, so photochemical decomposition plays an important role in their dissipation from soil (Gino, 1993). Dehalogenation and hydroxylation appear to be the predominant reactions eg: paraquat (1, 1-dimethyl – 4, 4 - bipyridillium dichloride) is converted to 4-carboxy, l- methyl pyridillium ion and methylamine (Funderburk and Bozarth, 1967). In phenoxy acids (2,4-D) the side chain is removed and becomes 2,4-dichlorophenol (Bell, 1965).
2. **Adsorption by soil particles:** Process of adsorption results in binding of a chemical to sites on soil minerals or organic surfaces. Adsorption is the key factor determining the environmental fate of a herbicide in soil, biological activity and persistence. Factors affecting adsorption of herbicides by soil are type of clay colloid e.g. inorganic colloids such

as montmorillonite with high CEC adsorb more. Other factors include soil organic matter content, pH, soil moisture and chemical nature of the herbicide. Herbicides like 2,4-D which are weakly acidic, releases H^+ ions in a neutral solution or basic solution and hence the anion is adsorbed readily at higher pH. Atrazine which is basic in nature is adsorbed to a greater extent in acidic soils (Harris and Warren, 1964). At higher pH, there is a chance of crop injury due to lesser adsorption of atrazine by soil which necessitates reduction in dosage.

A. High energy bonds

1. Ionic bonds: These bonds occur between organic anions or cations and have + ive or -ive electric charges located at the adsorbent surface. Some herbicide molecules are always in cationic form (paraquat, diquat etc.). While others can be ionised, ionisation depends on acidity of the medium. These compounds are either weak acids (eg. phenoxy acetic acids) or weak bases (e.g. triazines). Adsorption by ionic bonds is due to ion exchange (Knight and Denny, 1970).
2. Ligand exchange: Ligand exchange was postulated for the binding of 1,3,5-triazines onto the residual transition metals of humic acids (Hamaker and Thomson, 1972), for adsorption of linuron on clays (Hance, 1971), and for adsorption of terbutryne on aluminium saturated clays (Terce and Calvet, 1975).

B. Low energy bonds

Hydrogen bonding is a mechanism for the adsorption on montmorillonite of carbamates (Bailey et al, 1968) and atrazine (Calvet and Terce, 1975).

Variation in cumulative desorption between the soils may be due to the heterogeneity of soil with sorptive sites that vary widely in type and energy of bonding (Shanti et a1, 1997).

3. Chemical decomposition

In general, chemical degradation is not a significant pathway in herbicide decomposition. However, for herbicides like triazines, dinitroanilines, diphenyl ethers etc. chemical degradation is an important pathway for dissipation of the chemical from soil. Chemical decomposition involves chemical reactions like hydrolysis, oxidation, reduction, dealkylation etc. The rate of degradation is proportional to its concentration. It usually begins immediately after the herbicide is applied to the soil without any lag phase and the process continues at a steady state or declining state. Particles would create a zone of lower pH resulting in rapid acid hydrolysis.

C. Microbial transformations

Microorganisms occupy only a very small volume of the soil matrix (<0. 1%), however they are mainly responsible for numerous transformations that recycle elements and energy in nature. Microorganisms such as bacteria, fungi and actinomycetes are capable of detoxifying herbicides in soil. Among, these, bacteria predominates. During microbial degradation, herbicides exhibit varying degrees of inhibitory effects on soil microorganisms, reducing their population. The resistant organisms survive on the dead organic matter of the killed ones, and thus the final population consists of organisms tolerant to the particular herbicide.

Microbes degrade herbicides predominantly through ester or amide hydrolysis, alkylation, dealkylation, dehalogenation, reduction, oxidation, aromatic ring hydroxylation, ring cleavage and conjugation pathways. Therefore, microbial transformation is significant in the case of herbicides. Aliphatics, amides, arsenicals, benzoics, carbamates, nitriles, phenols, phenoxy acids are almost exclusively degraded by soil microorganisms. Chemical degradation is very little, thiocarbamates, trizaines, glyphosate etc. Diquat and paraquat (bipyridiliums) are strongly adsorbed in soils preventing microbial degradation. In case of dinitroanilines and triazoles, degradation by soil organisms accounts only for a small fraction of their total dissipation in the soil.

The adaptive mechanisms help microbes to develop capacity to degrade chemicals in the soil. Mineralisation of 2,4-D takes 2.5 months and atrazine one month (Willems et al, 1996).

Formation of Soil Bound Residues

Formation of bound residues in soil is a combination of various biotic and abiotic processes. Ecological consequences of this formation are contrary depending on the properties of the material to which the herbicide is bound. Binding of herbicides to high molecular insoluble organic material or to insoluble organic matter results in a drastic decrease of the mobility and bioavailability of the herbicide residue. This type of binding makes the herbicide unextractable and thus inaccessible to detection by usual analytical methods.These bound residues can be quantified by radiolabelled compounds. However, labeling technique do not provide any information about the chemical structure. Advanced instrumentation techniques viz. biological material oxidisers can be used for estimation of bound residues in various matrices.

Factors Affecting Persistence of Herbicides

Persistence of chemicals in soil is affected by physicochemical soil properties such as content of organic matter, kind of clay minerals, clay content, CEC, pH, moisture content, soil structure and soil texture. These properties affect the persistence of each herbicide in a different way.

Soil Moisture

Both chemical and microbial degradation depends on moisture content.

Mabbayad et al. (1986) found that faster dissipation of butachlor occurred at higher soil moisture and higher temperatures. Prakash and Devi (2000) reported that butachlor dissipation was faster under field capacity followed by submergence.

Soil pH

Soil pH has significant effect on detoxification of herbicides by influencing their ionic and molecular characteristics, ionic character and cation exchange capacity of soil colloids and the inherent capacity of soil micro organisms to react with the herbicides.

Three times more alachlor was adsorbed at pH 4.1 than at pH 12 in three soils with organic matter content less than 1.1 per cent (Sethi and Chopra, 1975). Acidification of soils greatly enhanced their ability to retain 2,4-D (Grover, 1977)

Soil Texture

Clay content is one of the principal factors influencing persistence of herbicides in soil, because clay and organic matter contents are themselves correlated. It Pesticides are deactivated more in the coarse fraction of soils compared with finer fractions. Glass (1986) found that adsorption of glyphosate by soils was related to clay content and CEC of soil. Cations such as Al^{3+} and Fe^{3+} can form hydroxide on the clay surface. This increases adsorption capacity of clay minerals (Calvet, 1980)

Soil Organic Matter

Soil organic matter is another important factor influencing adsorption of herbicides in soil (Karrie et al., 1991). Long term herbicide trials conducted at Kerala Agicultural University, Thrissur also showed similar results (KAU, 2003). Slowest degradation was observed in samples containing little organic matter with least adsorption (Hance, 1974). This was attributed to increased microbial activity in organic rich soils. Organic matter improved bioefficacy of herbicides in rice- rice system by increasing the absorption of herbicides.

Soil Microorganisms

Microorganisms are able to degrade a wide variety of chemicals, from simple polysaccharides, aminoacids, proteins, lipids etc. to more complex materials such as plant residues, waxes and rubber. Many herbicides are retained in the surface layer of soil for a relatively long term and are degraded mostly by soil microorganisms.

Walker et al. (1992) observed that degradation rate was positively correlated with microbial biomass. Both microbial population and adsorption were positively correlated with soil organic matter. Chakraborty and Bhattacharya (1991) found that two soil fungi namely *Fusarium solanum* and *Fusarium oxysporum,* effectively degraded butachlor in a 0.02M KH_2PO_4 buffer solution at pH 5.2.

Aspergillus sp. is the major organism responsible for degradation of 2,4-D in acid soils. It has clearly been shown that growth of *Aspergillus* sp.in the soil is enhanced by higher rates of application of 2,4-D in rice field

Herbicide Residues in Aquatic System

Herbicides are sprayed in the early stages of the crop, at the time when plant cover is little and hence major portion of the chemical reaches the soil where it becomes a part of the biological cycle in the environment. Potential sources of pollution of aquatic systems with herbicides are

(i) Herbicides applied for weed management in rice

(ii) Chemicals applied for aquatic weed control

(iii) Herbicides applied to field crops and plantations in hilly areas.

Among the three sources, herbicidal weed control in aquatic areas has more direct impact on aquatic system. The EEC Directive concerning the quality of water for human consumption established the maximum concentration of each pesticide at 0.1 µg/l and the total pesticide concentration at 0.5 g/l. (EEC, 1988; Vettorazzi, 1979).

Water solubility is the major factor determining the extent of pollution caused by a particular herbicide in ground water. Herbicides are considered as potential pollutants when the solubility is > 30 mg l^{-1}. Hydrolysis half life(> 25 weeks), photolysis half life(> 1 week), field dissipation half life (> 3 weeks), speciation(negatively charged, fully or partially at ambient pH) are the other factors that indicate high ground water contamination potential of herbicides. Herbicides coming under the groups viz., phenoxy alkanoics, sulfonyl ureas and bipyridilliums are more water soluble than others

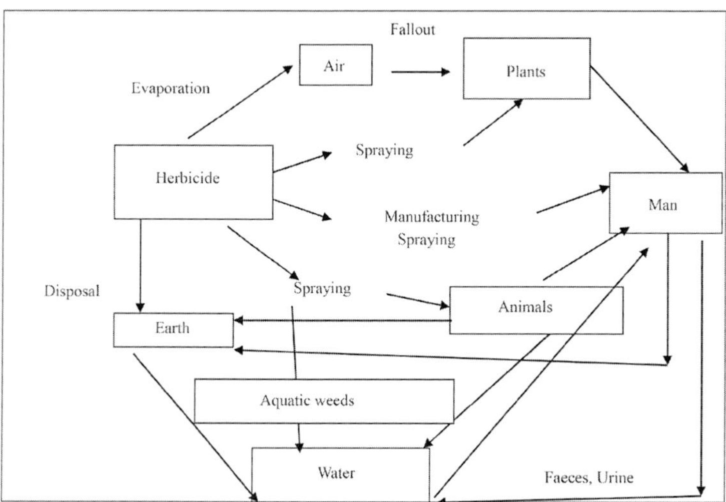

Fig. Herbicide cycle in the environment

Water Solubility of Important Groups of Herbicides

Herbicide group	Water solubility class
Dinitroanilines(eg.Pendimethalin-0.3 mg/l)	Very low (< 10 mg/l)
Thiocarbamates(eg.Thiobencarb-30 mg/l)	Low-moderate (10-1000 mg/l)
Chloroacetamides(eg.Butachlor-20 mg/l)	Low -Moderate (10-1000 mg/l)
Substituted ureas(eg.Diuron-42 mg/l)	Very low –low (10- 100 mg/l)
Phenoxy alkanoics(eg.2,4-D-900 mg/l)	Low-high(10- 10000 mg/l)
Sulfonylurea (eg.Chlorimuron ethyl-1200 mg/l)	Very low-high (pH dependent) (<10-10000 mg/l)
Organophosphates(eg.Glyphosate-10000 mg/l)	Very high(>10000 mg/l)
Triazines(eg.Atrazine-70 mg/l)	Low-moderate (10- 1000 mg/l)

Source: WSSA, 1994.

Recently, the impact of herbicides on drinking water and fish population has attracted much public attention. Many reports from different parts of the world indicated that fishes accumulate different concentration of pesticides that can enter the human body. Toxicity studies in fish reported from India have revealed significant accumulation of organo-chlorines in tissue. Therefore, recommendations on aquatic weed control can be evolved only if the residue studies confirm that the chemicals are quickly dissipated from all the components of the aquatic system. Common aquatic algaeicides and herbicides are given in

Toxicity of Common Aquatic Algaecides and Herbicides

Name of the compound	Rate of application	Target plants	Toxicity
A. Algaeicides 1) Copper sulphate pentahydrate	0.1 ppm 1.00 ppm 5.00 ppm	Waterbloom common scum algae *Chara* sp.	2.3 to 12.0 ppm in human drinking water. 100 ppm in animal drinking water
2) S. triazines Ametryn, Terbutryn Simazine	0.05 to 0.1 ppm 0.5 to 1.25 ppm		Unsafe for treating irrigation and drinking water bodies
3) Dichlone	0.02to 0.05 ppm 0.15 ppm	Water bloom Algal scums	Toxic to fish, treated water is unfit for drinking
4) Diuron	0.2 - 0.4 ppm	Macro algae	Accumulate in fish tissues
5) Hexazinone	0.5 ppm	Macro algae	Toxic to fish

B. Aquatic herbicides

1) Acrolein (Acrylaldehyde)	**15-80 ppm**	**Submersed weeds**	**Causes massive fish kill**
2) Amitrole and amitrole – T	5-15 kg ha^{-1} 1-1.5 kg ha^{-1}	*Typha* and *Phragmites* *Eichhornia crassipes*	Safe to fish, water is unsafe for drinking
3) Anhydrous ammonia	10 ppm	Submersed weeds (*Hydrilla, Najas* etc.) Waterbloom	Destroy all vegetations as well as fish and other fauna. However, soon after the weed control, safe for various uses.
4) Dichlobenil	5-10 kg ha^{-1} or 1 ppm	Emersed and Submersed macrophytes	Low toxicity to fish and wildlife.
5) Diquat and paraquat	0.25 to 1.5 ppm	*Submersed weeds Pistia, Lemna, Eichhornia, crassipes*	Unsafe for drinking, Diquat is safe to fish.
6) Diuron	10 kg ha^{-1} 1-2 ppm	Soil sterilant for treatment of water	
7) Fenac	10-20 kg ha^{-1}	*Potomogeton, Myriophyllum, Elodea, Najas* sp.	Unsuitable for drinking
8) Glyphosate	1-3 kg ha^{-1} in 300 l water	*Nymphaea, Phragmites communis*	Non residual
9) Silvex (fenoxaprop)	0.5 to 4.0 ppm	*Alteranthera philoxeroides, Nymphaea, Nelumbo*	Unfit for irrigation and other uses.
10) 2,4-D	1-8 kg ha^{-1} or 2 ppm	*Eichhornia crassipes, Myriophyllum, Najas* sp.	Unfit for irrigation, Toxicity to fish depends on the type of formulation used

Source: Gupta, 2001

References

Bailey, G.W., White, J.L. and Rothberg, T. 1968. Adsorption of organic herbicides by montmorillonite. Role of pH and chemical character by adsorbate. Proc. Soil Sci. Soc. Amer. 32:222-225

Bell, G.R. 1965. Photochemical degradation of 2.4-dichlorophenoxy acetic acid and structurally related compounds in the presence of riboflavin. Bot. Gaz. 118:113

Calvet, R. 1980. Adsorption-desorption phenomena. In: Hance, R.J. (ed.). Interactions between Herbicides and Soil. Academic Press, London, pp. 83-122

Calvet, R. and Terce, M. 1975. Adsorption of de l' atrazine par des montmorillonites. Al. Ann. Agron. 26: 693-707

Chakraborty, S. K. and Bhattacharya, A.1991. Degradation of butachlor by two soil fungi. Chemosphere 23(1): 99-105

Davies, J. 2001.Herbicide safeners-commercial products and tools for agrochemical research. Pesticide outlook. February: 11-15

Devi, K.M.D., Abraham, C.T. and Upasana C. N. 2015. Leaching behavior of four herbicides in two soils of Kerala . Indian J. Weed Sci.47(2):81-84

Edwards, C.A. 1973. Pesticides residues in soil and water. In: Edwards, C.A. (ed.). Environmental Pollution by Pesticides. Plenum Press, London, pp. 409-458

EEC [European Economic Community] 1988. Drinking Water Directive, Official Journal N 229/11, Directive 80/778/EEC

Funderburk, H.H. and Bozarth, G.A. 1967. Review of the metabolism and decomposition of diquat and paraquat J. Agr. Food Chem. 15:563-567

Gino, C. 1993. Influence of agronomic factors on the persistence of herbicides. In: Proc. Int. Symp. Indian Soc. Weed Science. Hisar, Nov. 18-20, 1993. Vol. I. pp. 121-128

Grover, R.1977. Mobility of dicamba, picloram and 2,4-D in soil columns. Weed Sci. 25: 159-162

Gupta, O.P. 2001. Weedy Aquatic Plants: Their Utility, Menace and Management. Agribios (India), Jodhpur, 273 p.

Hamaker, J.W. and Thompson, J.W. 1972. Adsorption. In: Goring, C.A.I. and Hamaker,_J.W (ed.) Organic Chemicals in the Soil Environment. Marcel Dekker, NewYork. pp. 49-144

Hamner, C.L. and Tukey, H.B. 1944. The herbicidal action of 2,4-D and 2, 4-5-T on bindweed. Science 100: 154-155

Hance, R.J. 1971. Complex formation as an adsorption mechanism for linuron and atrazine. Weed Res. 2: 106-110

Hance, R.J. 1974. Soil organic matter and the adsorption and decomposition of the herbicide atrazinc and linuron. Soil Biol. Biochem. 6: 39-42

Harris, C.I.and Warren, C.F. 1964. Adsorption and desorption of herbicides by soil. Weeds 12:120-126

Harris, C.R. 1972. Factors influencing the effectiveness of soil insecticides. A. Rev. Ent. 17: 177-198

Hartley, G.S. 1964. The physiology and biochemistry of herbicides. In: Audus, L.J. (ed.), Herbicides. Academic Press, New York, pp. 111-161

Kannan, M.M. 2003. Persistence of herbicides in rice-rice system M.Sc.thesis,Kerala Agricultural KAU , Thrissur , 195p.

Knight, B.A.G. and Denny, P.J. 1970. The interaction of paraquat with soil, adsorption by an expanding lattice clay mineral. Weed Res. 10: 40-48

Mabbayad, M.O., Moody, K. and Argenta, A.M. 1986. Effect of soil type, temperature and soil moisture on butachlor dissipation. Philipp. J. Weed Sci. 13: 1-10

Mallory –Smith C.A. and Retzinger, E. J. 2003. Classification of herbicides by site of action for weed resistance management strategies. Weed Technol. 17:605-619

Norris, LA. "The Movement, Persistence, and Fate of the Phenoxy Herbicides andTCDD in the Forest." Residue Reviews, 1981. 80, 65-135.

Pokorny, R. 1941. Some chlorophenoxyacetic acids. J. Am. Chem. Soc. 63: 1768

Prakash, N.B. and Devi, S.L. 2000. Persistence of butachlor in soils under different moisture regimes. J. Indian Soc. Soil Sci. 48(2): 249-256

Retzinger, E. J. and Mallory – Smith. 1997. Classification of herbicides by site of action for weed resistance management strategies. Weed Technol. 11:384-393

Rosinger , C. 2014. Herbicide safeners: an overview. 26th German conference on weed biology and weed control. March11-13, 2014, Braunschweig, Germany

Sankaran, S. Jayakumar, R. and Kempuchetty, N. 1993. Herbicide residues, Gandhi book house, Coimbatore, 251p.

Scheunert, I., Schroll, R. and Dorfler, U. 1993. Persistence of herbicides in Agricultural soils. In: Proc. Int. Symp. Indian Soc. Weed Science. Hisar, Nov. 18-20, 1993, Vol. I. pp. 96-98

Schmidt, R.R. 1998. Classification of herbicides according to mode of action. Leverkusen, Germany: Bayer Ag. 8p.

Sethi, R.K. and Chopra, S.L. 1975. Adsorption, degradation and leaching of alachlor in some soils. J. Indian Soc. Soil Sci. 23: 184-194

Shanti, G., Chandrasekhar, P. and Raman. S. 1997. Adsorption and desorption of isoproturon on eight soils of Hyderabad, Andhra Pradesh. J. Indian Soc. Soil Sci. 45(3):494-497

Stearman, G.K. and Wells, M. 1997. Leaching run off of Simazine, 2,4-D and Bromide from nursery plots. J. Soil Wat. Conserv. 52: 132-144

Stephenson , G. R. and Yacooby, T. 1991. Milestones in the development of herbicide safeners. Z.Naturforsch. 46c.: 794-797

Terce, M. and Calvet, R. 1975. Role of aluminium in the adsorption of atraine by clay minerals. In: Proceedings of the Israel France Symposium on Behaviour of Pesticides in soil. M. Horowitz (ed.) Bet Dagen, Israel, p.33-40

Vettorazzi, G. 1979. International Regulatory Aspects for Pesticide Chemicals Vol. II.CRC Press, Boca Raton. 141 p.

Walker, A., Moon, Y.H. and Welch, S.J. 1992. Influence of temperature, soil moisture and soil characteristics on the persistence of alachlor. Pestic. Sci. 35(2): 109-116

Willems, H.P.L. Lewis, K.J., Dyson, J.S. and Lewis, F.J. 1996. Mineralization of 2, 4-D and atrazine in the unsaturated zone of a sandy loam soil. Soil Biol. Biochem. 28: 989-996

WSSA, 1994. Herbicide Handbook , Weed Science Society of America: Champaign

Zimmerman, P.W. and Wilcoxin, F.L. 1935. Several chemical growth substances which cause initiation of roots and other responses in plants. Contrib. Boyce Thompson Inst. 7: 209-229

13

Aquatic Weeds and Their Management

A major portion of the earth's surface is water. Apart from the oceans, major water bodies on the earth's surface include rivers, streams, lakes, ponds and canals. Plants adapted to these ecosystems form the aquatic species. When they pose a problem to navigation, fisheries and agriculture and also cause health hazards, they are categorised as aquatic weeds. Problems due to aquatic weeds has been reported from different parts of the world. They are common in warm waters. In Florida, it is reported that water hyacinth (Eichhornia *crassipes*) covers nearly 40,000 hectares of water resources. In Bangladesh 300t/ha of the weed is carried by flood water which destroys rice fields. In Phillipines it is reported to form a mat in 90,000 ha of lake Laguna de Bay and affect pisciculture. In Srilanka, *Salvinia* sp. covers 12000 ha of swamp and rice fields.

Many wet land ecosystems which remain inundated during rainy season are also infested with aquatic weeds. Their removal and destruction before land preparation is a time consuming costly affair.

Water transport is comparatively cheaper and eco friendly method to transport goods and services in areas where canals and rivers are abundant. Alien Invasive species such as Salvinia and Eichhornia have become a major impediment throughout the world as they have completely invaded our canals and rivers.

Aquatic weeds can be defined as those plants which grow and complete their life cycle in water and cause harm to the aquatic system directly and to related eco environment relatively (Lancar and Krake, 2002). Aquatic plants are a natural part of the aquatic system. They are required to maintain normal health and productivity of the aquatic environment but when they become too abundant in the ecosystem they compete with fishes for oxygen and nutrients, clog water ways and cause environmental pollution. These plants are not desired by managers of water bodies when they grow in abundance and interfere with the growth of other crop plants or ornamentals in the ecosystem (Pieterse, 1990). Many introduced alien plant species have become a threat to the aquatic ecosystem in many parts of the world.

These introduced plant species have spread to large areas in a short span of time because of their invasive nature, and have become serious problems. As most of the aquatic areas are no man's lands and not owned by individuals, their spread is not initially noticed or controlled. Hence controlling large stretches of aquatic areas has become a challenge to governmental agencies and this causes a drain on the exchequer. It is also important to find environmentally safe, non- chemical management approaches. Awareness of the type of aquatic weeds, problems caused by aquatic weeds, their biology, and available control methods is essential for managing these weeds and to restore the health of the ecosystem.

Problems from Aquatic Weeds

Excessive growth of aquatic weeds reduces the storage area of dams, ponds and tanks. Reports indicate that 50 km of Bheema river track in Maharashtra is blocked every year due to rampant growth of *P. stratiotes*. This causes great problem for taking water from the river for irrigation purposes..

"Water blooms" are due to the rapid multiplication of some plankton species which form dense masses when environmental conditions and availability of nutrients are favourable. These dense growth also imparts colour to water bodies like green, reddish brown, yellow green and blue-green depending on type of algae. Temporary blooms are due to Chlorophyceae (for example, *Chlamydomonas* spp., *Pandorina morum, Volvox aureus, Chlorella vulgaris*), Bacillariophyceae (*Melosira granulata, Synedra ulna*), Dinophyceae (*Peridinium inconspicuum*) and Euglenineae (*Euglena* spp., *Trachelomonas* spp.). Permanent blooms are caused by Myxophyceae (*Microcystis* spp., *Anabaena* spp., *Raphidiopsis* spp., *Oscillatoria chlorina*). Common filamentous algae *Spirogyra, Pithophora* and *Oedogonium* can also bloom but their presence gives a green tinge to water.

Thick growth of aquatic weeds in canals and irrigation channels reduces water flow. They also interfere with hydroelectric schemes by creating problems to turbines. Problems have been reported from several hydro electric projects of India like Nagarjuna Sagar in Andrapradesh, Tunkabadra project in Karnataka, Kakki and Idukki projects in Kerala which have been suffering from massive growth of aquatic weeds (Sushilkumar, 2011).

Due to excessive transpiration, loss of water occurs from reservoirs. Water loss by transpiration through leaves of weeds from a surface fully covered by aquatic weeds is found to be much higher than the loss through evaporation from a water surface free of weeds.

Aquatic weeds cause hindrance to navigation by blocking water ways and choking propellers of motorized boats. This has been reported from places where goods and services are transported by water. Kuttanad area of Kerala is a region from where such problems are constantly reported..

Sushilkumar et al., 2009 reported aquatic species such as *A caroliniana, Polygonum* sp., *Cyperus* sp. and *A philoxeroides* as problematic weeds in low lying areas in Jabalpur. These species blocked drainage canals of colonies and irrigation canals by their profuse growth.

Psiculture area in India is estimated to be 8 lakh ha, 40% of this is rendered unsuitable by aquatic weeds. Most of the fishery tanks and ponds in and around Bangalore and other cities have been badly invaded by water hyacinth. Aquatic plants like *Eichhornia, Azolla, Nymphaea, Nelumbo, Nymphoides, Hydrilla, Vallisneria, Potamogeton, Najas, Ceratophyllum, Typha* and *Utricularia* spp. are problematic weeds in fishery lakes and tanks of Andhra Pradesh, Assam, Haryana, Himachal Pradesh, Jammu & Kashmir, Maharashtra, Tamil Nadu and Uttar Pradesh in India.

Problems due to water hyacinth has been reported from Utter Pradesh. The Keetham lake near Agra had an annual fish production of Rs. 2 Lakh. This lake had an area of 300 ha. Pisciculture in the lake was reduced to almost nil by aquatic weeds.

Floating aquatic weeds affect the quality of water and also reduce the aesthetic value of the land scape. This adversely affect recreation in water bodies

Increase of mosquitoes cause health hazards. In low lying and inundated areas such as Allapuzha district of Kerala water born diseases and diseases caused by mosquitoes such as Dengu fever are a common occurrence.

Table 1: Important aquatic weeds in different states of India

State	Weed species	Water body
Assam, Orissa, West Bengal	*Azolla pinnata, Chara* spp. *E. crassipes, Ceratophyllum* spp., *H. verticillata, Ipomoea aquatica, Lemna minor, Monochoria vaginalis, Marsilia quadrifolia, Nymphaea* spp, *Nitella* spp., *Sagittaria* spp., *Scirpus* spp., *Nelumbo* spp., *Pistia* spp., *Najas* spp., *Ipomea carnea, Trapa bispinosa, Salvinia molesta*	Fisheries, ponds, tanks, water works, deep water rice and lakes.

State	Weed species	Water body
Andhra Pradesh, Kerala, Karnataka, Tamil Nadu	*E. crassipes, Cyperus* spp, *Chara* spp.,*Ipomoea aqatica, H. verticillata, Nymphaea* spp, *Nelumbo lutea; Nymphoides* spp. *Potamogeton* spp., *Najas* spp., *Salvinia molesta, Typha latifolia, Vallisnaria americana.*	Lakes and tanks growing fishes, irrigation and drainage systems.
Punjab, Bihar, Uttar Pradesh, Haryana,	*Cyperus aquatica, E. crassipes, Hydrilla verticilla, Ipomoea carnea, Najas* spp. *V. spirallis, P. nodosus, Nelumbo nucifera, Nymphaea* spp., *Phragmites karka, Chara* spp. *Potamogeton crispus, T. latifolia, Ceratophyllum, Typha angustata; P. zosterifolius, P. perfoliatus, P. pectinatus, Spirogyra* spp. *Myriophyllum spicatum, green algae Vallisnaria americana*	Irrigation canals, drainage system, ponds, lakes, fisheries areas and rivers.
Gujarat, Rajasthan, Madhya Pradesh	*Chara* spp., *E. crassipes Hydrilla verticillata,Typha latifolia Ipomoea carnea, I. aquatica, P. nodosus, Vallisnaria americana, Phragmites karka,V. spirallis, Potamogeton crispus,Nymphoides* spp.	Water storage reservoirs for city water supply system, fisheries development, irrigation canals and drainage system.
Jammu and Kashmir.	*Nymphoides peltatum, Trapa natans Potamogeton* spp., *Salvinia natans, Spirodela polyrhiza, Lemna gibba. Polygonum amphibium, L. trisuleha, L. minor,*	Natural water-bodies for storage, aquatic sports and aesthetic value, Lakes

Reproduced from Sushilkumar (2011)

Classification of Aquatic Plants

Aquatic plants can be classified into five groups (Denny, 1985), as follows;

1. Free floating weeds: Free floating with most of the stem, leaf or tissue at or above the water surface. E.g. *Eichhornia crassipes*, *Salvinia molesta*, *Pistia stratiotes*, *Lemna minor*, *Azolla pinnata.*

2. Emergent/emersed weeds: Rooted plants with most of their leaf and stem tissues above the water surface E.g. *Phragmites karka*, *Typha angustata*, *Alternanthera philoxeroides*, *Ipomoea fistulosa*, *Limnocharis flava*, *Ipomoea aquatica*, *Ludwigia adscendens*.

3. Rooted weeds with floating leaves: Rooted plants with most of the leaf tissues at water surface. E.g. *Nymphaea nouchalli*, *Nelumbo nucifera*

4. Submerged weeds: Most of the vegetative tissue beneath the water surface; they are rooted or attached to the bottom of the water body by root like organs. E.g. *Hydrilla verticillata*, *Potamogeton pectinatus*, *Elodea canadensis*, *Vallisneria torta*, *Cabomba furcata*.
5. Algae: Unicellular or filamentous lower plants without differentiated tissues which grow at, or below, the water surface. Eg. *Microcystis spirogyra*, *Hydrodictyon* spp.

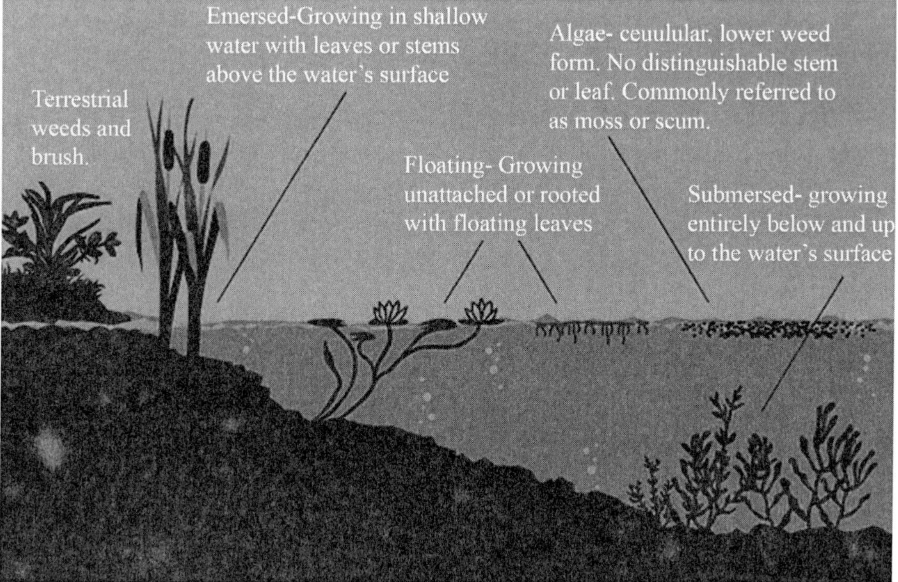

Fig 1: Classification of aquatic weeds (reproduced from Invas Biosecurity-Ireland)

Problem Aquatic Weeds

1. *Eichhornia crassipes* (Mart.) Solms

Common names: Water hyacinth, Pickerel weed, Lilac devil

Family: Pontederiaceae

Origin: Brazil

E. crassipes is a weed with worldwide distribution in lakes, rivers, ponds and ditches It is named after J.A.F. Eichhorn (1779-1856), the Prussian minister of Education and Public welfare, court advisor and politician. The species name is derived from Latin 'crass' meaning 'thick', indicating that it is a plant with thick leaf stalks.

It is believed to have been introduced to the US in 1884 at an exposition in New Orleans. Within 70 years of reaching Florida, the plant covered 1, 24,000 acres of water ways (Schmitz et al, 1993). It was introduced to the Indian sub continent from Malaysia or Indonesia during 1890 from tropical South America (Cook, 1996). Currently over 5 lakh ha is infested by water hyacinth. In India, a large amount of money is spent for removing it from the major rivers like the Ganges. It is a perennial (occasionally annual) monocot flowering plant, usually floating (except when stranded in the mud). The mother plants and daughter plants are attached by floating stolons. Leaves are formed in rosettes rising to three feet above water, and are entire, ovate, rounded or broadly elliptic, upto 6 inches wide, thick, glossy, waxy green and water proof. Petioles have inflated bulbous spongy structures helping to float in water. Roots hang submerged beneath floating leaves, dark purplish or black in colour, feathery with long root caps at the tip. It produces flowers with blue perianth (8-15 numbers) in a single, very showy spike (about 12 inches long) at the top of an erect, thick stalk rising above the leaves. The fruit is a 3 celled capsule producing numerous oblong seeds of 1.5 mm size.

E. crassipes is distributed globally in the tropics and subtropics, but it's spread is limited by severe cold conditions. It grows faster than any other tested plant when the temperature range is between 20-30^0 C (Kasselmann, 1995). Temperatures above 35^0 C and below 12^0 C reduces growth rate (Wolverton and McDonald, 1979) . Plant population is reported to double within six days (Mitchell, 1976) depending on the organic content of water. It can reproduce vegetatively and sexually. Vegetatively, new rosettes (daughter plants) are formed on floating stolons which grow from the mother plants.

Eichhornia mats clog waterways making boating, fishing and almost all other activities impossible, as water flow through the hyacinth mats is greatly diminished. Water hyacinth mats degrade the water quality by blocking the air-water interface and greatly reducing the oxygen levels in the water, eliminating under water animals such as fish. The weed reduces the biological diversity by eliminating naturally submerged plants by blocking sunlight, and alters emersed plant communities by pushing away and crushing them.

E. crassipes is utilized locally as fodder, manure, a source of methane and alcohol, for purifying water, for making handicraft mats and baskets and is also grown as a decorative aquarium plant.

2. *Salvinia molesta* Mitchell.

Common name:Water fern, Kariba weed, African payal (in Kerala)

Family:Salviniaceae

Origin: Native of south eastern Brazil (Forno and Harley, 1979)

The genus name Salvinia is given after Italian scholar Antonia Maria Salvinii (1653-1729). The genus has 13 species, many of which are distributed only in South America. Up to 1973, *S. molesta* was misidentified as *Salvinia auriculata* Aubl. Studies conducted by Dr. D.S. Mitchell have revealed that it is a distinct taxon, different from *S. auriculata* (Joy, 1978).

S. molesta flourishes from sea level to 900 m. Mitchell (1972) suggested that *S.molesta* is a hybrid involving *S. auriculata* and *S. biloba* or *S. herzogii*, all from South America.

Salvinia might have spread to Asia and Africa through botanical gardens. Specimens of Salvinia were brought from Germany to Royal Botanical Gardens in Calcutta in 1933 and from there living material was dispatched to Colombo University in Ceylon (Sri Lanka) for research purposes. Unwanted specimens were expelled into nearby waters and they multiplied and established beyond all expectations (Joy, 1978). From Sri Lanka it reached Kerala (Cook and Gutt, 1971) where it became a serious menace. It invaded large areas in backwaters of Kerala severely affecting the water transport along the canals in Alleppey district. The thick growth of Salvinia in paddy fields had to be removed before land preparation, expending huge amounts for labour charges. Ultimately it was brought under control by releasing the weevil *Cyrtobagous salviniae* for biological control (Joy, 1978).

Salvinia is a floating plant with horizontal stem, immersed below water surface, with pairs of floating leaves arising at intervals. Root like structures grow out into water from each node. Upper surface of the leaves bear rows of non wettable multi cellular papillae. Plant is covered by wettable, uniseriate, multicellular hairs. As the plant is sterile reproduction is only by vegetative means, broken pieces of mother plant developing into new ones. Plant favours open condition with good sunlight, water rich in nutrients, moderate temperature of about 20^0 C and protection from wind and water currents for growth. It is an aggressive weed, which has invaded most of the tropical and subtropical regions of the world (Harley and Ferno, 1990).

3. *Marsilea* spp.

Family: Marsileaceae

Common name: Four leaf clover, European water clover

There are 45 species of Marsilea distributed in the tropical and warm regions of the world. Stems creeping or floating, rooting at nodes. Leaves compound with four leaflets arranged symmetrically cross wise, each obovate, margin entire when growing in water and becoming toothed when dry. Sporangia are closed, sporocarps borne on short stalks arising from the petioles or entire axils. Spores of two kinds are produced, usually found together in the same sporocarp; megaspores solitary, microspores numerous. The sporocarps may remain viable for 100 years. They may also germinate after passing through a duck. Sporocarps are dispersed by birds (endozoochory), other animals, by water and probably also in mud.

The plant is seen in temperate Eurasia and naturalized in North America. In India it is mainly observed in NW Himalayas. Three major species have been reported from India.

1. *M. aegyptica* Wildenow: a perennial usually found in shallow water in seasonally flooded areas.
2. *M. coromandelina* Wildenow: a delicate plant usually found at edges of small pools and buffalo wallows and in seasonally swampy areas.
3. *M. minuta* Linnaeus: an extremely variable plant, cushion like on dry land, but spreading and creeping in water. Found in shallow pools, at the edges of rivers, canals and ditches and in rice fields.

Marselia is considered to be a troublesome weed in rice fields and irrigation ditches. Used locally in medicine. In rice fields where 2, 4-D has been continuously used for years, other dicot weeds were controlled, while Marselia resistant to 2,4-D became a menace. Almix® (Chlorimuron ethyl 10% + Metsulfuron methyl 10%) is found to be effective in controlling Marselia in rice fields.

4. *Alternanthera philoxeroides* (Martius) Grisebach.

Common name: Alligator weed

Family:Amaranthaceae

Origin: South America

Distributed in North America, Asia and Australia. A perennial herb, stem often forming tangled masses in water (often partially floating) or along the shore. It is considered to be a very noxious weed. In China, it is regarded as a valuable food for livestock particularly pigs, used as a green manure, vegetable and in medicine (Majumdar and Banarjee, 1976). It is a fast spreading weed in aquatic areas in India.

5. *Pistia stratiotes* (L.)

Family: Araceae

Common name: Water lettuce

Origin: Tropics and sub tropics of America

Now naturalized in India and South East Asia. It is a free floating plant with rosettes of densely pubescent leaves and numerous roots. Perennial or annual, usually seen in fresh water bodies but sometimes in brackish or even salt water. Propagated by berries, seeds or detached rosettes, mostly dispersed by water but fruits may be eaten by animals and seeds spread.

6. *Ipomoea aquatica* Forsskal

Family: Convolvulaceae

Common name: Water spinach

Commonly seen in India, Bangladesh, South East Asia, Taiwan and Southern China. A perennial weed floating in stagnant water but sometimes found on the banks of pools, canals and rivers. It is often cultivated for its edible shoots and as a medicine. It can be a serious weed if left unchecked.

7. *Ipomoea fistulosa* Martius *ex* Choicy *(I. carnea)*

Family: Convolvulaceae

Common name: Shoe flower, Gramaphone plant

Origin: South America

Naturalized in Africa and Asia.

Usually a perennial shrub of 2.5 m or more height, found on dry rocks, in places subject to flooding, in permanent swamps or in water upto 2 m depth. It grows in fresh and brackish water. It has become a serious weed in irrigation and drainage channels (Cook, 1987) and is seen throughout India along shallow, submersible areas and ponds. Thick growth of *I. fistulosa* is a problem on the shorelines of dams and lakes, reducing the water storage area in addition to loss of water through transpiration .

8. *Eleocharis* spp.

Family: Cyperaceae

Common name:Spikerush

Annual or perennial plants often with creeping rhizomes or stolons. Most of them occur in shallow water, some persist submerged in deeper water. Many are reported as weeds but some are cultivated for matting, as decorative aquarium plants or as food.

Eleocharis dulcis (N.L. Burman) Trinius *ex* Henschel

Common name: Chinese water chestnut

Origin: Old world tropics

Perennial in nature, gregarious in shallow water, in ponds, rice fields and along irrigation canals. Sometimes it grows in salt or brackish water. It is a wide spread and very variable species but immediately recognized by the culms with transverse septa with spikelets narrower than the culms. A cultivar, Tuberosa, is cultivated for the edible tubers and sold under the name Chinese water chestnut.

9. *Hydrilla verticillata* (L.f.)Royle

Family: Hydrocharitaceae

Common name: Hydrilla, water thyme

Originated in the old world. It is one of the most notorious submerged aquatic plants forming masses growing in still or slowly flowing water, often very abundant and dominant in large areas. It is considered to be a serious weed in North America. In India it is a particularly troublesome in tanks used for growing fish. It is a submersed perennial herb, rooted, with long stems that branch at the water surface where growth becomes horizontal and forms dense mats. Small, pointed leaves are arranged in whorls of four to eight. Flowers are tiny, white and grow on long stalks. Produce ¼ to ½ inch long, off white to yellowish, potato like tubers attached to the roots. Thick mat like growth of Hydrilla block sunlight penetration to native plants growing below. Hydrilla seriously affects water flow and water use. Its heavy growth may obstruct boating, swimming and fishing in lakes and rivers and block withdrawal of water used for power generation and agricultural use. Growth of fish is adversely affected and fishing in shallow water becomes impossible.

Control of Hydrilla is achieved by use of lime, herbicides (Copper sulphate, Diquat, Endothal and Furidon) and biological control by using grass carp or by mechanical removal. For mechanical control hand cutting tools and aquatic plant harvesting equipments are usually used. Rotovators are also used for underwater tillage and removal of the root crowns and tubers to effect long term control. Use of bottom barriers like geotextile ground cover cloths or erosion control materials attached to the bottom by pins or sandbags has been found to be effective for management of pioneering infestations of Hydrilla and preventing its spread.

10. *Limnocharis flava (L.)*

Family: Pontederiaceae

Common name: Water cabbage, Yellow velvet leaf

Originated in tropical America. Introduced to Malaysia in 1870 and recently in Bangladesh and India. Perennial or annual plant growing in shallow water, drying mud and in floating mats of vegetation. Recently large populations have been seen in the rice fields and uncultivated waste lands of backwater regions of Kerala. It is spreading fast and becoming a problem weed of rice. It is used as a fodder and as fresh vegetable. If incorporated into the soil, it is a good green manure. Herbicides like 2, 4 – D, Almix® (Chlorimuron ethyl + Metsulfuron methyl) are effective for control of Limnocharis without causing any phytotoxicity to rice.

11. *Polygonum pulchrum* Blume (*P. tomentosum* Willdenow)

Common name: Smart weed

Family: Polygonaceae

It is a perennial shrub with erect stems growing more than 1 m tall, common on sides of irrigation canals, rivers, and aquatic waste areas . It is seen often encroaching paddy fields left fallow for years.

12. *Nymphaea* spp. (Water lily) and *Nelumbo* spp. (Lotus)

These are rooted aquatic plants producing large roundish leaves floating flat on the water surface. Both water lily and lotus produce highly attractive flowers borne well above the water on long peduncles. Unripe seeds of lotus are edible and rhizomes are used as vegetable. However, in many situations, infestation of water lily and lotus creates obstructions for movement of water, boats and other activities. Leaves cover the water surface and affect the growth of fishes.

Two species of *Nymphaea* commonly seen in India are *N.nouchali* N.I. Burman [*N. stellate* Willdenow] and *N. pubescens* Willdenow [*N. rubra* Roxburgh ex Andrews].

In *N. nouchali* the leaf blade is entire to bluntly dentate and glabrous while in *N. pubescens* leaf blades are sharply dentate and pubescent below (hairs sometimes restricted to veins)

In lotus the leaves are round and the petiole is attached in the middle of the lower surface whereas in *Nymphaea* the leaves are deeply cut at the base.

13. *Typha domingensis* Persoon

(*T. angustifolia* L.)

Common name: Cattail

Family:Typhaceae

Perennial aquatic weed found in a variety of habitats. Juvenile plants are submerged while adult plants are emergent or sometimes terrestrial. It is usually gregarious and often dominant over large areas. Stems erect and corm-like, connected by stolons or rhizomes.

14. *Monochoria vaginalis (Burm.f.) Presl ex Kunth.*

Common name: Pickerel weed, Marshy betel vine

Family: Pontederiaceae

Annual or sometimes perennial, growing in swamps, edges of pools, ditches and canals. A common weed in wetland rice weeds. It is highly plastic and has stems from 3 cm to over 50 cm tall. In fertile, organic fields the growth is gregarious and may be mistaken for *Eichhornia*. Plant is rooted in the mud and its upper portion grows above water. It has succulent stems and shiny ovate-acuminate leaves. Flowers are bluish purple in axillary racemes. It reproduces by seeds.

15. *Cabomba furcata* Schult. & Schult.f.

Common name: Red Cabomba, Forked fanwort

Family: Cabombaceae

Cabomba furcata is an aquatic perennial herbaceous plant native to S. America. It grows rooted in the mud of stagnant to flowing water. It is a submersed, sometimes floating, but is considered as a rooted fresh water plant. It bears purple flowers and is used as an aquarium plant. The plant requires good sunlight and flourishes in open areas like canals, shallow ponds and flooded paddy fields. The plant has feathery lime green to deep red forked leaves. In soft acidic waters it grows as dense mat. This plant is a problem in many parts of India.

16. *Utricularia vulgaris* (L.)

Common name: Common bladderwort, Greater bladderwort

Family: Utriculariaceae

Free floating, does not produce roots. Stems can attain over 1 m length. Leaves are finely pinnately divided and carry many bladder like traps. Yellow flowers are borne on stalks above the surface of the water. A related species, *U. macrorhiza,* is a large perennial carnivorous plant.

17. *Ludwigia adscendens* (L.) Hara

Common name: Creeping water primrose

Family: Onagraceae

A creeping or floating herb on water, with spongy floats and rooting at the nodes. The plant is occasionally seen along the margins of ponds, water channels and rice fields. The plant produces white flowers with yellowish tinge at the center.

18. *Phrag mites karka* (Retz.) Trin. *ex* Steud

Common name: Common reed, Water reed

Family: Poaceae

It is a large perennial grass found in wetlands of tropical regions. It forms extensive stands on the margins of aquatic bodies like canals and lakes. It is spread by horizontal runners which produce roots and shoots at regular intervals. The erect stems grow 2 to 6 meters tall. The plant is a halophyte and tolerates brackish water.

19. *Nymphoides indica* (L.) Kuntze

Common name: Water snowflake, Floating hearts

Family: Menyanthaceae

A floating weed of shallow wetlands, *Nymphoides indica* is easily distinguished by flat heart shaped floating leaves. It is a perennial water plant with the blooms looking like snowflakes on water. Flowers are white with yellow centers. Petals have unusual feathery edges. New plants are formed all the time where the floating stolons form tufted plantlets along their lengths. The mother plant has a short, thick stem which is rooted in the mud at the bottom of the water body

20. *Limnophila heterophylla* Benth.

Common name: Marshweed

Family: Scrophulariaceae

Limnophila heterophylla is an emergent annual herb with highly dissected submerged leaves and opposite or verticillate, serrated aerial leaves, it is a typical example of heterophylly. The plant grows in and around lowland ponds, streams and rice fields. Flowers are formed in the upper axils. Dense growth can impede the flow of water in irrigation channels.

Management of Aquatic Weeds

To control aquatic weeds, physical, chemical and biological methods are used depending on different situations.

A. Physical methods

Manual techniques: Most common method adopted for aquatic weed control. These includes use of manual labour for hand pulling, raking, cutting, collecting etc. practiced in underdeveloped countries where labour is comparatively cheap. Manual collection and heaping of floating aquatic weeds like *Eichhornia* and *Salvinia* before land preparation and sowing rice is a common practice in the lowland paddy fields of Kerala, India. Main drawbacks for manual methods is high cost and lack of complete control of all weeds. A few weeds left in the field will multiply fast and cover the field soon, unless follow up weeding is practiced. Varshney and Singh (1976) considered that, in India, manual weeding was partially successful in 65 to 90 per cent cases for control of *Nelumbo nucifera, Pistia stratiotes, Nymphae astellata* and *Hydrilla verticillata.*

B. Mechanical methods

A large number of diverse machines are available to cut, shred, crush and suck or roll aquatic weeds.

Table 1: Common methods of mechanical removal of aquatic weeds

Method	Operation	Remarks
1. Dredging	Removal of sediments and soil with weeds and their propagules	For canals, shoreline areas (after draining or drying), or by using machines mounted on floating structures like pontoons in wet situations
2. Weed cutter/ harvester	Cut and remove weeds to a desired depth Harvesting of floating weeds	For removal of top growth of weeds. Re -growth could occur from the stumps, requiring repeated operations. JCB machine mounted floating rafts to cut and remove floating mats of weeds like water hyacinth.
3. Suction	Collection of floating weeds like *Salvinia* and *Eichhornia*	Machine mounted on a raft collects crushes and disposes the weeds on the banks.

These machines are useful for controlling weeds in canals and rivers. But for extensive areas (like lakes) they have limitations. It is costly and maintenance of the specific machinery exclusively for weed control is difficult. Moreover, weeds cut down or harvested by the machinery have to be removed from the aquatic body, to prevent contamination of water. In irrigation canals and other small areas, mechanical removal is beneficial as it has the potential for quick and predictable removal of weeds from specified locations which can be achieved rapidly. Hand cutting tools are often used to control aquatic weed infestations in small sections of lake or river systems. For controlling infestations in large areas, aquatic harvesters are used which will cut, harvest, crush and deliver weeds on the shore line. Rotovators are used to penetrate the soil, till it and release the crowns and other plant tissues. They can till the bottom sediments upto a depth of 20 feet and remove the root crowns below sediment structures of the weeds. Materials like geotextile ground cover cloth and erosion control materials are laid across sections of lake and river bottom to control limited infestations of noxious weeds and to prevent their further spread. These barriers are attached to the bottom by using pins or sand bags to give long term control. Diver dredge technology is also used to control pioneering infestations of weeds like Hydrilla. In diver dredging operations, divers use venturi pump systems to collect plant and root biomass. Pumps are mounted on barges or pontoon boats, each craft can support two dredge systems. Dredge hoes are from 3 to 5 inches in diameter and are handled underwater by one diver. Diver dredging can provide excellent results for control of limited infestations of weeds like Hydrilla. This technology can remove the plant and tubers from the lake system and prevent the spread of weeds. In Kerala, JCB machines are used for clearing weeds from irrigation canals in the low lying Kole lands in Thrissur district.

C. Chemical methods

Herbicides offer cheap, effective and rapid control of aquatic weeds. They are powerful tools which require knowledge and experience for safe and effective use. If misused, they can have side effects which may be harmful to aquatic organisms. Herbicides may be selective (e.g. dalapon which controls grasses but not broad leaf weeds), or non selective (e.g. glyphosate which control all green plants). Contact herbicides (eg. diquat) kill only those parts of the plant on which they fall (usually the foliage), but if sufficient damage is caused, the whole plant may die. Translocated herbicides (e.g. glyphosate) are absorbed by one part of the plant but move within the plant and act on other tissues or growing points. Persistent herbicides (e.g. fluridone) retain their activity in soil or water for some time, usually measured in weeks or months. Non-persistent herbicides (e.g. glyphosate) act only when sprayed directly onto foliage and lose their phytotoxic activity very quickly on contact with soil water.

A list of the common herbicides used for aquatic weed control is given in the Table2. Herbicide application varies with the type of weed. For controlling emergent and floating weeds, usually herbicides are sprayed just like any terrestrial application. Submerged weeds and algae are controlled by applying herbicide to water so that it is absorbed by submerged foliage or roots. Application rate is usually calculated on the basis of volume of water being treated rather than the area of the water surface.

Table 2: Important herbicides used in aquatic weed control

Herbicide	Type of weeds controlled	Dose (ml/L or kg/ha)
Sodium arsenate	Submerged weeds	5-8 ml/L
Copper sulphate	Submerged weeds and algae	0.5-2 mg/L
Hydrogen peroxide	Submerged weeds	10-20 mg/L
Dalapon	Emerged grass weeds	18-25 kg/ha
2,4-D	Free floating and emerged weeds	2-10 kg/ha
	Submerged weeds	1 mg/L
Dichlobenil	Submerged and emersed floating weeds	1-2 mg/L
Diuron	Algae, floating and emerged weeds	0.5-1.5 mg/L
Triazines	Algae, submerged and floating weeds	0.05-1 mg/L
Paraquat	Floating weeds and emergent weeds	0.8 – 1.6 kg/ha
Diquat	Algae, submerged weeds and floating weeds	0.5-1 mg/L (submerged weeds)
		1.0 kg/ha (floating and emergent weeds)
Endothal	Submerged weeds	0.5-2.5 mg / L water
Fluridone	Submerged and floating weeds	0.1-1 mg/L
Glyphosate	Emergent and floating weeds and marginal weeds.	1.8-2.1 kg/ha

Among the above herbicides, 2,4-D, paraquat and glyphosate are used in India for control of floating and emergent weeds. Copper sulphate and its mixture with 2,4-D are usually recommended for control of algal growth in ponds.

Diquat dibromide and endothal are two of the oldest and most commonly used herbicides in Australia and New Zealand for aquatic weed control. Diquat is a broad spectrum herbicide effective on free floating weeds including salvinia, water hyacinth, azolla and water lettuce, and it is often mixed with copper based products to control submerged aquatic plants including algae. Endothal is primarily used to control submerged aquatic plants, especially hydrilla, against which diquat is not effective. To increase the efficacy of diquat and endothal, various carrier adjuvants are mixed, such as alginate gum, guar gum and methocel. The most widely used gel adjuvant is Aquagel® marketed as Hydrogel® in Australia. It is made of guar gum which is a non toxic polysaccharide starch. It has been found that by addition of the upto 50 % less herbicide usage will give the same result as when the herbicide is used alone (Chisholm et al., 2015).

Chemical control of weeds at the time of land preparation for rice

Many of the low lying rice fields in the backwater areas are located at elevations below mean sea level. Only one crop of rice is taken in these areas when water level recedes during summer season. For most part of the year, the fields are flooded and are covered with aquatic weeds, mainly the emersed and floating type. At the time of land preparation, water is pumped from the field and the weeds are gathered and heaped manually. This operation is a costly affair as it may require large number of labourers, sometimes upto 150 per hectare. Because of the high cost of labour and the non-availability during peak period of demand, many farmers leave the land fallow. Trials conducted in Kerala showed that by spraying glyphosate (Roundup ® 7.5-10 L/ha), or paraquat (Gramoxone® 5L /ha), these weeds can be dried by two weeks after spraying. If some weeds escape the herbicide spray (can be seen as green patches), a second spray can be given to these areas. Glyphosate was found more effective against water hyacinth and grass weeds, while paraquat was effective against *Salvinia*. Once the weeds have become brown due to the action of herbicide, the field can be ploughed with tractor to incorporate the weeds in to soil and puddle the field. By this the expenses for collecting and heaping the weeds can be reduced. Moreover, growth and yield of the crop is found to be better due to supply of nutrients from the decomposed weeds. The study also indicated that incorporation of weeds and puddling of the field by tractor is not necessary, once the weeds are killed by the herbicides, sprouted rice seeds can be broadcast sown without any further land preparation. The fields under 'zero tillage' gives yields on par with the puddled fields.

Limitations of chemical weed control

Use of chemicals for control of aquatic vegetation has its impact on the ecology of the aquatic systems due to (1) direct toxicity to both target and non-target organisms and (2) indirect effect associated with the destruction of target macrophytes. In addition to effects on target plants in affected area, residues of the herbicides applied may affect the plants in the non target area by spray drift or residue movement in water currents and sediments. Chemicals may have adverse effects on fishes and other organisms in water, directly due to their toxicity. Indirectly, decaying of dead target plants may lead to increase in CO_2 concentration, decrease in dissolved O_2 levels and changes in pH of water. Movement of herbicide through water current and infiltration may lead to residue problems in drinking water. Therefore, widespread use of herbicides in large areas should not be undertaken without assessing the possible ecological side effects.

Effect of residues applied for control of aquatic weeds is always a concern. In a laboratory study conducted under AICRP on Weed Control of Kerala

Agricultural University, Devi et al. (2011) found residues of paraquat and 2, 4 –D to be below detectable levels by 60 days after spraying and their residues in fish samples were below acceptable daily intake level of 0 .0004 and 0.01 mg per kg body weight respectively. 2,4 – D caused greater adverse effect on liver and muscle fibers of fishes than paraquat or glyphosate.

D. Biological methods

Biological control of aquatic weeds could be defined as 'activities aimed at decreasing the population of an aquatic weed to acceptable levels by means of a living organism'. This can be achieved by using selective organisms which attack one or only a few species. Selective control of many problem aquatic weeds was successful with many arthropods and a few fungi. Fishes are the main non- selective organisms used for aquatic weed control, although other organisms like snails and manatee were also found useful in some areas. Competition for critical factors for growth, such as light, can be used for suppression of aquatic weeds. Planting trees and shrubs along side streams and ditches is reported to reduce aquatic weed problems (Pieterse and Zon, 1982). A list of the successful bio-control agents against some of the problem weeds of aquatic areas are given in Table 3.

Table 3: Organisms used for biological control of aquatic weeds

Type of organism	Name of the weed	Biocontrol organism
Arthropods	*Alternanthera philoxeroides* (Alligator weed)	*Agasicles hygrophila* (Coleoptera: Chrysomelidae)
	Salvinia molesta (Salvinia)	1.*Cyrtobagous salviniae* & *C. singularis* (Coleoptera: Curculionidae) 2. *Paulinia acuminata* (Orthoptera: Acrididae) 3. *Samea multiplicalis* (Lepidoptera: Pyralidae)
	Eichhornia crassipes (Water hyacinth)	1. *Neochetina eichhorniae* and *N. bruchi* (Coleoptera: Curculionidae) 2. *Sameodes albiguttalis* (Lepidoptera: Pyralidae) 3. *Orthogalumna terebrantis* (Acarnia: Galumnidae)
	Pistia stratiotes (Water lettuce)	1. *Neohydronomous pulchellus* (Coleoptera: Curculionidae) 2. *Epipasmmia pectinicornis* (Lepidoptera: Noctuidae)
	Hydrilla verticillata (Hydrilla)	1. *Parapoynx dimunutalis* (Lepidoptera: Pyralidae) 2. *Bagous* sp. (Coleoptera: Curculionidae) 3. *Hydrella spp* (Diptera: Ephydridae)

Type of organism	Name of the weed	Biocontrol organism
Fungi	*Eichhornia crassipes*	1. *Alternaria alternata* 2. *A. eichhorniae* 3. *Cercospora rodmanii* 4. *Fusariumequisetii*
	Hydrilla verticillata	*Fusariumroseum*
	Pistia stratiotes	1. *Cercospora* sp. 2. *Sclerotium rolfsii*
	Salvinia molesta	*Myrothecium roridum*
	Alternanthera philoxeroides	*Alternanthera alternantherae*
Phytophagous fish	Different aquatic weeds	1. *Ctenopharyngodon idella* (Grass carp) 2. *Hypothalmichthys molitrix* (Silver carp) 3. *Tilapia melanopleurea* (Tilapia)

Integrated control of aquatic weeds

Integrated control is the management system that judiciously utilizes all suitable techniques to reduce pest populations and maintain them at levels below those causing economic injury. Integrated use of herbicides and a biological agent was found to be successful in many areas. Some of the successful cases are summarized below:

Table 4: Successful cases of biological control of weeds

Name of the weed	Successful integrated approach	Location
Alligator weed	2,4-D + Flea beetle	USA
Water hyacinth	2,4-D + Water hyacinth weevil	USA
	Cercospora rodmanii + Water hyacinth weevil	USA
	Grass carp + water hyacinth weevil	USA
	Grass carp + mech. means (cut down the weed during spring and early summer)	USA
Hydrilla verticillata and Myriophyllum spicatum	Reduce water level and give 2,4-D spray on exposed weed and diquat application on portions still covered by water	

Integrated control of aquatic weeds is still in experimental stage. In most circumstances, weed control activities are taken up only when weed vegetation has become troublesome to optimal use of a water body. At this stage, a single method which gives quick effect is resorted to. However, now with the availability of computer based management systems to assist in management of complex fresh water systems such as irrigation networks, integrated management practices can be planned and implemented.

Case Studies

1. Salvinia molesta

One of the spectacular achievements of biological control of an aquatic weed in India is the control of *Salvinia molesta* in Kerala. The aquatic fern, *Salvinia molesta* Mitchell, of Brazilian origin, had become a very serious problem in the backwater areas of Kerala during the 1960s and 70s. The weed was characterised by abundant growth and seemed formidable to the available methods of control.

A project on biological control of Salvinia was initiated by the Kerala Agricultural University in 1972, which later merged with the All India Co-ordinated Research Project on Biological Control of Crop Pests and Weeds (ICAR) in 1977. Attempts to control the weed by introducing the grass hopper *Paulinia acuminata* De Geer through the CIBC (Commonwealth Institute of Biological Control), Bangalore centre, were not successful as the multiplication of the grasshopper in the field was drastically retarded by the attack of predators like spiders.

In 1982, a consignment of the weevil *Cyrtobagous salviniae* Calder and Sands was obtained from Australia. After quarantine studies, field release of the weevil was carried out throughout Kerala. *Cyrtobagous salviniae* established in most areas where it was released and Salvinia was brought under control within 12 to 14 months. Salvinia choked canals are not seen anymore. About 2000 sq. km area of the weed has been cleared by *Cyrtobagous salviniae* (Joy et al., 1987). The most commendable aspect of successful control of Salvinia in Kerala is that it involved only negligible expenditure, as once the weevil was released, it multiplied on the weed and got spread by the movement of the weed along with the tidal flow of water.

First instar larvae of *Cyrtobagous salviniae* scrape the green matter on young leaf buds producing linear scars on tender leaves. After four to five days, the larvae enter the plant tissue through the basal portion of leaves and either remain there or tunnel at the nodal region of rhizome and this area turns brownish after about a week. The leaves then become characteristically darkened, resulting in death and disintegration of plant tissues. Thus, major damage is caused by tunneling of rhizome by the larvae. Adults feed on unopened leaf buds and tender leaves. Because of their feeding, circular holes are formed on the buds and leaves. The weevils occasionally feed on rhizomes, producing scars, and on roots, causing disintegration at the point of attack. Visible symptoms of attack by *Cyrtobagous salviniae* include gradual change of colour from normal green to yellowish green, and then to rusty brown. Leaf size and root length get gradually reduced and finally the weed turns into a brownish black mass.

2. *Eichhornia crassipes*

Water hyacinth is native to Brazil and other central and South American countries, but has spread to most tropical and subtropical regions of the world where it has become a very serious weed. Research on biological control began in 1961 and control agents were first released about ten years later.

Weevils *Neochetina eichhorniae* and *N. bruchi* were the most successful bio-agents. Adult weevils damage leaves and larvae damage plants by tunneling towards the base of the petiole and into the crown, where they cause severe damage. Between the two weevils *N. eichhorniae* is more effective.

Larvae of the moth *Sameodes albiguttalis* feed inside the petioles and buds, and attack may be heavy but sporadic as the moth has a preference for tender, often bulbous petioles.

Biological control of *E. crassipes* infestation was very successful in Sudan in the White Nile System, where the spread of the weed had badly affected the commercial steamer traffic and access to the river side villages, in addition to increased water loss through evapotranspiration. *N. eichhorniae* was released in 1978, *N.bruchi*in 1979 and *S. albiguttalis*in 1980. All three insects established well.

Eichhornia plants throughout the system were seen scarred and growth of the weed was suppressed by the insects. It took about five years for *N. Eichhorniae* to give noticeable reduction in the weed infestation.

These bioagents have been introduced to many other countries where *E.crassipes* has become a problem. Throughout the introduced range of *E. crassipes*, the potential for biological control is excellent. Among the three insects, *N. eichhorniae* is the most promising one

Cercospora rodmanii is a fungus which causes small brown black spots on leaves and petioles. As a result of these spots, the leaves die back from the tip to the base of the petiole until the entire leaf is killed. The plants with severe infections become chlorotic and dark brown. Eventually the leaves die and sink (Freeman and Charudattan, 1984). This type of disease progression occurs only under controlled conditions during a period of several weeks to months, under severe and sustained disease pressure. Under natural field conditions severity is much less. However, the growth of the weed is drastically reduced.

Host range tests with 85 plants in 22 families showed that only water hyacinth is highly susceptible to *C. rodmanii*. Field tests during 1974 to 1984 in several locations in Florida showed that the pathogen established well, and within three weeks after application of inoculum the fungus was capable of secondary spread by wind-borne spores.

Use of *C. rodmanii* for biological control of water hyacinth was patented by the University of Florida, and Abbott Laboratories USA was licensed to develop the fungus as a microbial herbicide for commercial use.

Studies conducted on the combined action of *C. rodmanii* and *Neochetina* sp. gave very encouraging results. Combination of the fungus and the insects was capable of eliminating water hyacinth from the test plots by seven months, while the fungus alone or the insects alone could only get reduction in the growth rate of the weed.

Uses of Weeds

Phytoremediation by aquatic weeds

Rapid industrialization and urbanization leads to discharge of large amounts of sewage effluents and municipal wastes into aquatic bodies. These wastes are rich in organic matter, plant nutrients, organic and inorganic materials and heavy metals. Heavy metals like cadmium, lead, mercury, arsenic, etc cannot be destroyed or changed to harmless forms. Even at low concentrations, these heavy metals are toxic to living organisms. They get into the food chains and increase health problems. Nutrients such as manganese, copper and zinc, though essential for plants at low concentrations, are toxic at high levels. Removal of heavy metals and contaminants from industrial wastes is done by several physico-chemical methods which are expensive and not satisfactorily effective. Search for simple and economical procedures have pointed to phytoremediation as a promising technique. Here the pollutants are removed by absorption and accumulation by plants.

Many aquatic weeds have the capacity to absorb high levels of metal contaminants. Metal accumulating plant species, referred to as hyper accumulators, can absorb and store significant amounts of metal contaminants in different parts of the plant. These plants are usually seen growing luxuriantly in areas of high pollution like sewage drains and in areas where factory effluents are discharged. Weeds like *Eichhornia crassipes* and *Alternanthera philoxeroides* have been proved to absorb heavy metals like mercury, silver, cobalt and strontium. Phytoremediation is recognized as a cost effective method for remediating sites contaminated with toxic metals, radio nuclides and hazardous organics at a cost lower than with conventional technologies.

In trials conducted at the Water Technology Centre, Indian Agricultural Research Institute (IARI), Delhi, *Typha*, *Acorus* and *Phragmites* were found to be very efficient in removal of pollutants from municipal waste water. Wetland treated waste waters had reduced turbidity, and lower content of chromium, lead, nickel and phosphate, (Khankhane and Kaur, 2015).

Vermicomposting of aquatic weeds

Aquatic weeds can be profitably utilized for producing high quality vermicompost which can be used for improving soil quality and crop yields. A study conducted by Girija et al. (2005) showed that *Salvinia molesta* and *Eichhornia crassipes* can be used for making vermicompost after mixing with rice straw, coconut leaves and cowdung slurr. Quality of compost produced was high in the case of *E. crassipes*, both in terms of nutritive value and recovery percentage of compost.

Prameela et al. (2011) reported that vermicomposting of alligator weed (*Alternanthera philoxeroides*) along with cowdung, banana pseudostem and coconut fronds for 90 days produced good quality compost with appreciable N, P and K contents.

Other uses of weeds

Aquatic weeds have many uses which vary with the locality and availability. Some of the common uses listed below.

Table 5: Common uses of aquatic weeds

Sl.No.	Use	Weed
1	Animal food	*Eichhornia crassipes, Alternanthera philoxeroides, Azolla, Ipomoea aquatica, Eleocharis dulcis*
2	Biofertilizer	*Azolla*
3	Compost	*Eichhornia crassipes, Pistia,Salvinia*
4	Biogas production	*Eichhornia crassipes, Hydrilla, Salvinia, Typha*
5	Biomass as mulch and organic manure	*Eichhornia crassipes, Hydrilla, Phragmites, Typha*
6	Human food	*Colocasia* sp., *Ipomoea aquatica, Eleocharis tuberosa*
7	Medicinal uses	*Acorus, Bacopa monnieri, Nelumbo, Nymphaea*
8	Mat and basket making	*Typha, Eichhornia crassipes*
8	Paper, pulp, fibre board	*Eichhornia crassipes, Pandanus, Phragmites, Typha, Eleocharis*
9	Thatching	*Phragmites, Typha*
10	Waste water treatment	*Eichhornia crassipes, Typha, Phragmites*

References

Charudattan, R. 1990. Pathogens with potential for weed control. ACS Symposium Series 1990, No. 439. p. 132-154

Chisholm, W., Chandrasena, N. and Harper, P. 2015. New herbicide application techniques for the management of aquatic weeds in Australasia. (Eds. B.N. Rao and N.T. Yaduraju) Weed Science for Sustainable Agriculture, Environment and Biodiversity, Vol. I. Proceedings of the Plenary and Lead Papers of XXV Asian-Pacific Weed Science Society Conference, Hyderabad, India. p. 283-293.

Cook, C.D.K. 1987. Ipomoea fistulosa - a new problem for India. Aquaphyte 7:12.

Cook, C.D.K. 1996. Aquatic and Wetland Plants of India. Oxford University Press Inc., New York. 385 p.

Cook, C.D.K. and Gutt, B.J., 1971. Salvinia in the state of Kerala, India.Pestie. Abstr.17:438-447

Denny, P. 1985. The ecology and management of African wet land vegetation. W. Junk, The Hague. 344 p.

Devi, K.M.D., Maya, S., Prasanna, K.S., Lalithakunjamma, C.R. and Abraham, C.T. 2011. Herbicide residues in the aquatic system and its impact on fish. Proceedings of the International Conference on Biodiversity and Aquatic Toxicology, February 12 – 14, 2011, department of Zoology and Aquaculture, AcharyaNagarjuna University, Nagarjuna Nagar. P. 104-111

Forno. T. W. and Harley, K.L.S. 1979. The occurrence of Salviniamolesta in Brazil. Aquatic Bot. 6:185-187.

Freeman, T.E. and Charudattan, R. 1984. Cercospo rarodmanii Conway – a biocontrol agent for waer hyacinth. Tech. Bull. 842, Agric. Exp. Stn. Sta., Univ. Florida, Gainsville.18 p.

Girija, T., Sushama, P.K. and Abraham, C.T. 2005.Vermicomposting of aquatic weeds. Ind. J. Weed Sci. 37(1 &2): 155-156

Harley, K.L.S. and Forno, I.W. 1990. Biological control of aquatic weeds by means of arthropods. Aquatic weeds: the ecology and management of nuisance aquatic vegetation. (Eds. A.H. Pieterse and K.J. Murphy) Oxford University Press, New York. pp. 177-186

Joy, P.J., Satheesan, N. V., Lyla, K.R. and Joseph, D. 1985. Biological control of weeds in Kerala.Proc. Nat. Symp. Entomophilic Insects, Calicut. p. 247-251

Joy, P.J., Satheesan, N.V. and Lyla, K.R. 1987. A friendly weevil.Intensive Agric. XXV (5-6): 12

Joy, P.J.1978. Ecology and control of Salvinia (African payal): the molesting weed of Kerala. Tech. Bull No.2.Kerala Agricultural University, Directorate of Extension Education, KAU, Thrissur. 40p.

Kasselmann, C. 1995. Aquarienpflanzen.Egen Ulmer GMBH & Co., Stuttgart. 472 pp. (In German)

Khankhane, P.J. and Kaur, R. 2015. Weeds as phytoremediation agents of aquatic ecosystem. Ind. Fmg. 65 (7): 63-65

Kumar M and Singh J. 1996b. Management of aquatic weeds in irrigation system. 28-35. In: Proceedings of Workshop on "Aquatic Weed-Problem and Management" (Ed. Varma, CVJ) held at Bangalore (Karnataka), 5-7 June 1996.

Majumdar, N.C and Banarjee, R.N. 1976.The distribution and uses of Alternantheraphiloxeroides. Bulletin of the Botanical Society of Bengal 30(1-2):147-148.

Mitchell, D.S., 1972. The Kariba weed, Salviniamolesta. British Fern Gazzette.10:251-252

Mitchell, D.S. 1976. The growth and management of Eichhorniacrassipes and Salvinia spp. in their natural environment and in alien situations. Aquatic weeds in South East Asia (eds. C. K.Varshney and J. Rzoska) Dr. W. Junk b.v., Publishers. The Hague: 396 p.

Pieterse, A.H. 1990. Concepts, ecology and characteristics of aquatic weeds – Introduction to aquatic weeds: The ecology and management of nuisance aquatic vegetation (eds. Pieterse, A.H. and K.J. Murphy) Oxford University Press, New York. pp. 3-16.

Pieterse A.H. and van Zon, J.C.J. 1983. Geintegreerde bestrijding van wateronkruiden in Egypte. In Geintegreerde bestrijding in de derde wereld, Wageningen, The Netharlands. pp. 111-19.

Prameela, P., Menon, M.V. and Sushama, P.K. 2011. Converting a menace to manure – vermicomposting of Alligator weed (Alternanthera philoxeroides). Abstracts of National Symposium on Waste Managememnt: Experiences and Strategies, 5-7 January 2011, Kerala Agricultural University, Thrissur. p. 32.

Schmitz, D.C, Schardt, J.D, Leslie, A.G., Dray, F.A, Osborne, J.A. and Nelson, B.V.1993. The ecological impact and management history of three invasive alien aquatic plants in Florida.Biological pollution- the control and impact of invasive exotic species. (ed. B.N. McKnight) Indiana Acad.Science. Indianapolis. 261 p.

Sushilkumar .2011.Aquatic weeds problems and management in India. Indian Journal of Weed Science 43 (3&4): 118-138,

Varshney, C.K. and Singh, K.P.1976. A survey of the aquatic weed problem in India. (Eds. C.K. Varshney and J. Rzoska) Aquatic weeds in South East Asia. W. Junk, The Hague, pp. 31-42.

Varshney Jay G, Sushilkumar and Mishra J.S. 2008..Current status of aquatic weeds and their management in India. Proceedings of Tal: (Eds. Sengupta M. and Dalwani R.) The 12th World Lake Conference, Jiapur, 28 October - 2nd November 2007: 1039- 1045.

Wolverton, B.C and Mc Donald, R.C, 1979. Water hyacinth (Eichhorniacrassipes) productivity and harvesting studies.Econ. Botany 33:1-10

World Bank Report. 2010. The World Bank Agriculture & Rural Development Data - data. worldbank.org/topic/agriculture- and-rural-development.

14

Parasitic Weeds and Their Control

Parasitic weeds are a major constraint in tropical agriculture. These are plants that are adapted to live on other plants sharing their nutrients and water. They constitute 1% of flowering plants, approximately 3000 - 5000 species coming under 20 families. Parasitic weeds from five families viz. Orobanchaceae, Convolvulaceae, Santalaceae, Scrophulariaceae and Loranthaceae are considered troublesome weeds as they infest economically important crops in many parts of the world and are very difficult to control by normal weed management measures. Mentioned below are some parasitic species belonging to these families and the crops infested by them

Table 1: Major families of parasitic weeds and their host plant

Family	Weed	Crops affected
Convolvulaceae	*Cuscuta campestris* *C. chinensis, C.reflexa*	Broadbean,Tomato, Niger, Linseed, Lentil, Onion Chickpea, Greengram, Blackgram, Lucerne, Sugarbeet etc.
Loranthaceae	*Dendrophthoe falcata* *Helicnthus elastica* *Helixanthera* spp	Mango, Jack fruit, Cocoa, Rubber, Teak, Tea,Coffee
Viscaceae	*Viscum album* *V. angulatum*	Forest trees
Orobanchaceae	*Orobanche crenata* *O. aegyptiaca,* *O. ramosa*	Sunflower, Tomato, Tobacco, Peas, Carrot, Beans, Potato, Watermelon, Mustard
Santalaceae	*Thesium* spp	Sugarcane, Onion
Scrophulariaceae	*Alectra* spp *Buchnera hispida,* *Striga* spp. (9 spp.)	Cowpea, Groundnut, Maize, Millet, Rice Sugarcane, Sorghum

Reproduced from Parasitic weeds biology and management (2014)

Parasitic weeds belong to two different categories. Some completely depend on host plant for their survival these are the complete parasite or holoparasites. When plants are capable of photosynthesis and depend on host only for support, nutrients and water they are partial parasites or heteroparasites. Development of haustoria, which forms a bridge between the host and the parasite, is the main feature of all parasitic plants. The term haustorium comes from the Latin word, *haurire* which means to drink. There are also some stem parasites which live completely within the stem of the host. These are endo parasites. The only part that emerges from the host is a tiny bud that opens into minute flowers e.g. Species belonging to the genera Apondanthes and Pilostyles have been reported from countries like Australia, Africa and Iran.

Among these, most troublesome weeds for field crops are witchweed or *Striga*, broomrape or *Orobanche* and dodder or *Cuscuta* (Parker and Riches, 1993). On tree crops in tropics, most important parasites are tropical mistletoes generally cited as Loranthus. It is generally observed that problems due to parasitic weed infestation are higher in underdeveloped countries where seed quality I sacrificed due to low purchasing power of farmers.

Broom Rape (*Orobanche*)

Broom rape is a native of the Mediterranean region (North Africa, Middle East, Southern Europe), also seen in Western Asia. It is a complete root parasite. Around 200 species have been reported over a wide range of host plants. It is an annual, host specific parasite mostly propagated through seed. In India it is reported to cause 50% loss in tobacco and around 80% loss in solanaceous vegetables (Punia, 2014). Most commonly seen *Orobanche* species in India and their host are given below.

Commonly seen *Orobanche* species and their hosts

Species	Host plant
Orobanche aegyptiaca Pers. Egyptian broomrape	Tomato, poatato, tobacco, brinjal, pea, fababean, cabbage, cauliflower, rape, mustard, turnip ,hemp, sunflower, spinach
Orobanche ramosa L. Branched or Hemp broomrape	Hemp, potato, tobacco, tomato, groundnut, cowpea, chilli, pepper
Orobanche crenata Forsk Bean broomrape	Fababean, peas, lentil, chickpea, vetch, clover
Orobanche cernua Loefl. Nodding broorape	Solanaceous crops, Tomato, potato, tobacco, brinjal
Orobanche Cumana Sunflower broomrape	Sunflower and tomato
Orobanche minor Sm. Common broomrape	Red and white clover, lucerne, tobacco, carrot, lettuce, sunflower and many others
Orobanche solmsii Changbao broomrape	Mustard and tobacco in Nepal

The plant is a prolific seed producer, reported to produce 250,000 seeds per plant. Seeds are minute and easily spread from one field to the other by water, wind, animals and man. Seeds retain viability even after passing through the alimentary canal of animals and remain viable in the soil upto 15 years. The life cycle of the weed is given in fig 1.

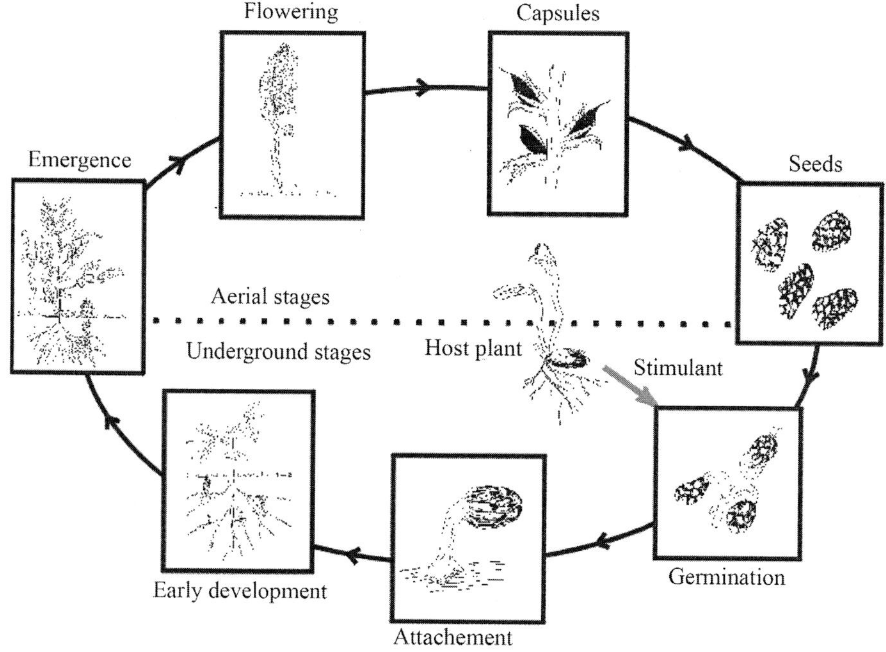

Life cycle of orobanche

(Adopted from uni-hohenheim.de/www380/380b/science/supraregional/CourseHH.htm)

Orobanche seeds have been reported to have staggered dormancy, while one half of the fully matured seeds remain viable the rest remain dormant. Seeds require preconditioning for germination for which a moist environment and suitable temperature for several days are pre requisites. Different species have varying optimum temperature range for germination and development. The conditioned seeds germinate in presence of a chemical stimulant released by the germinating host plant. Pre conditioned *Orobanche* seeds germinate only when a host root is nearby.

Control

Different approaches are used to control *Orobanche* in the field. The first and foremost aspect is preventing parasitic weed seed movement from field to field through irrigation water, farm implements and animals. Hand weeding reduces seed bank in the soil.

Summer ploughing and soil solarisation using clear polyethylene covers for 4–8 weeks and also flooding is recommended in areas of heavy infestation.

Growing trap crops like jowar, gingelly, blackgram and greengram and catch crops like berseem, sweet pepper are recommended to reduce weed infestation. Other control measures include changing sowing time and also using resistant genotype. Application of nitrogen with phosphorus and potassium was reported to reduce infestation (Abu-Irmaileh,1981).

Chemical control measures

Direct application of Strigol-GR-24 or GR-27 at 0.3kg/ha in acid soil to 1.5kg/ha in alkaline soil about 6 weeks before sowing the host plant to induce suicidal germination of the parasite (Saghir.1973).Other chemical treatments include drenching of plant holes with CuSO4, 5% solution, Use of pre-emergence herbicides such as Pendimethalin 0.75 to1.0kg/ha. Metolachlor or Alachlor 1.0kg/ha for solanaceous crops, pulses and oil seeds is reported to delay emergence of parasite. Basal application of neem cake at 150 to 200kg/ha. in rows lowers density of the parasite.

Witch Weed (*Striga*)

Striga spp. was earlier included in family Scrophulariaceae but currently its under Orobanchaceae. It is a seed propagated annual partial parasite. It is widely distributed. There are 18 to79 species, of these only 13 species are seen in tropical countries. *Striga* is a problem in most commercial field crops like maize, sorghum, peal millet, finger millet, sugarcane and even upland rice. Yield loss due to *Striga* infestation is recorded as 20-80% in Asia and Africa.

Seeds of Striga remain viable in the soil for many years. Stimulants released by host plants such as kinetin, zeatin, strigol, ethylene etc. activates growth of the parasitic seeds that have been conditioned after undergoing a period of dormancy. These seeds germinate and through hypha like cells form a bridge like connection with the host xylem. They depend on the host for water and mineral nutrients. Damage is observed in the vegetative growth of the plant. Host plant shows symptoms of wilting, stunted growth and ultimately it shrivels and dies or becomes unproductive.

Striga hermonthica, S. asiatica and *S.gesnerioides* are the most widely distributed species seen in India, Indonesia, Philippines, Africa and USA. Loss due to *Striga* infestation is reported to be greater than the combined loss due to all fungal diseases and insect pests except stalk –borer in South Africa (Saunders,1933). Severe infestation has been reported from many Indian states such as Maharastra Karnataka, Andhrapradesh and Tamilnadu, mainly in Sorghum. Life cycle of the weed is given in Fig.2

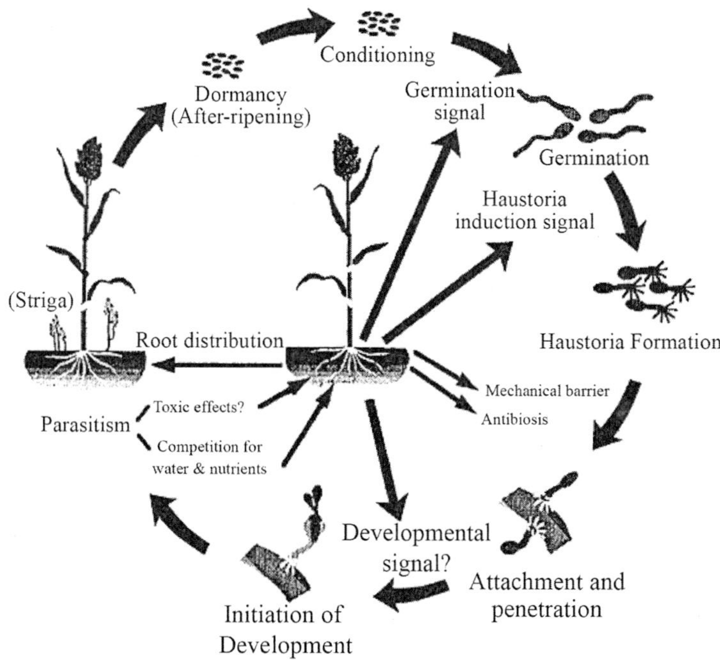

Fig. 2. Life cycle of Striga spp.
(Adopted from Ejeta and Butler, 1993. African Crop Science Journal)

Control

Use of clean and certified seed is the foremost preventive measure to be followed. Deep ploughing and incorporating seeds well below root zone helps to prevent the stimulant reaching seeds of *Striga*.

Soil solarisation, flooding, crop rotation, mulching, as well as improving soil fertility through manures and fertilizers will help to reduce density of infestation. Using resistant varieties and transplanting seedlings of maize, sorghum etc. is reported to be beneficial. Growing intercrops like silver leaf (*Desmodium uncinatum*), Sun hemp, groundnut, cowpea, soybean and red gram along with susceptible crops like maize, sorghum is reported to lower *Striga* infestation.

Direct application of herbicides such as 2,4-D at 1.0 to 2.0 kg/ha is reported to be a very practical alternative. Pre emergence application of oxyfluorfen 0.09-

0.12kg/ha. is also effective. Exogenous application of ethylene has been found to be effective in infested areas of USA.(Chancellor et al.1971)

Fusarium oxysporum f.sp *striga*, is a fungus identified as a potential biocontrol agent . It is reported to affect the weed at all stages of growth. Currently in use only in West Africa.

Dodder (*Cuscuta*)

Cuscuta, a parasitic angiosperm native to North America is identified as one of the most damaging parasitic weed worldwide. Adaptability of the weed to wide range of climatic conditions from warm temperate to sub-tropical and tropical condition has contributed to its wide geographical distribution. Moreover it has a wide host range and control is difficult. Earlier the weed was included in the family Convolvulaceae. In recent publications it is mentioned as Cuscutaceae. Genus *Cuscuta* has 175 species. In India 12 species have been reported (Gaur,1999). It is a serious problem in oilseeds, pulses, rice, ornamentals and fodder crops

Seeds of *Cuscuta* germinate near the soil surface. Seedling is rootless, leafless stem. After emergence they twin around the leaf or stem of the host. They establish connection to the host through haustoria. Once attached it remains there till harvest. Though *Cuscuta* contains chlorophyll it is an obligate parasite which cannot complete its life cycle without a host. Mainly propagated through seeds and to a lesser extend through shoot fragments.

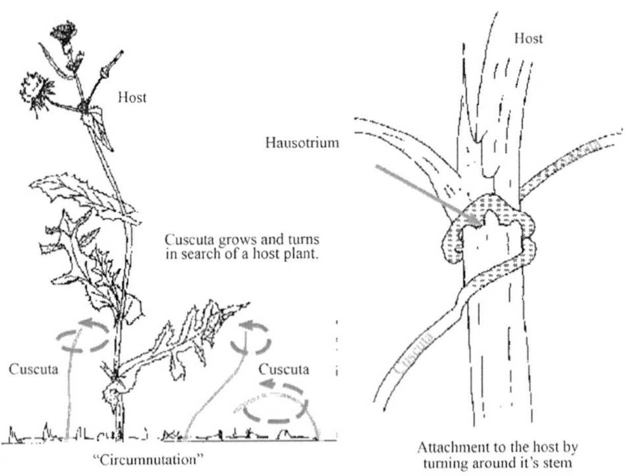

Fig. 3: Germination and attachment of *Cuscuta* spp.

Most common species in India are *Cuscuta campestris* Yuncker and *Cuscuta reflexa* Roxb.

Cuscuta campestris Yuncker

It is the most widespread species reported from Madhya Pradesh, Gujarat, Andhra Pradesh, Orissa. It is a major pest on field crops like alfalfa, niger, blackgram, greengram, lentil, chickpea and linseed (Mishra, 2009).

Cuscuta reflexa Roxb.

It is a hardy species mainly seen on woody plants and shrubs. The robust vine is 1-2mm thick and has higher level of chlorophyll. Mostly seen in Hyderabad region also reported from Nepal, Pakistan, Sri Lanka and Afghanistan.

Cassytha

This is parasitic weed species very similar to *Cuscuta* but belongs to the family Lauraceae. It is also known as laurel dodder or love vine. There are 20 species under this genus

Among these, *Cassytha filiformis* L. is very common in India. *Cuscuta reflexa* is very similar to this species that it is difficult to distinguish them. The main difference is in the seed character. Seeds of *Cuscuta* is a capsule while that of *Cassitha* is a drupe.

Both *Cuscuta* and *Cassytha* species are reported to have many medicinal properties. The plant is used to strengthen kidney and liver, is an anti- aging, anti - inflammatory agent, pain reliever and used for treatment of head ache. It is also recommended for decreased eye sight, vertigo chronic diarrhoea. *Cassytha* extract is used for curing skin diseases, cleaning ulcers and also for dysentery.

Control

Avoid using crop seeds from infected areas. It is possible to separate *Cuscuta* seeds from the alfalfa seed by mixing the alfalfa seed with iron filings and then exposing them to electromagnets. Similarly velvet covered rollers are found to capture the dodder seeds from the seed lot .This is possible because of the rough surface of the dodder seeds. So equipping the seed processing industry with facilities for removing *Cuscuta* seeds will help to reduce dodder problem in endemic areas.

Adopting stale seed bed technique after land preparation will help to eliminate seeds from soil seed bank. Crop rotation using cereals and forage grass will help to reduce density of the weed by 90%. Use of tolerant varieties is suggested in endemic areas.

Pre-emergence herbicides trifluralin or pendimethalin at 0.75-1.50kg/ha is reported to lower the menace of *Cuscuta*. Post- emergence spray (10 days after sowing) of pedimethalin at 0.50kg/ha was observed to effectively kill the emerging *Cuscuta* in crops like lucerne (Trivedi et al 2000 and Chinnusamy, 2014). Contact herbicides and systemic herbicides like glyphosate are also found to be effective on the weed.

Tropical Mistletoes (Tree parasites)

All hemi parasitic plants belonging to order Santales are known as Mistletoes. There are five families in this order of which plants belonging to Loranthaceae, Viscaceae and Santalaceae are major tree parasites. Members of Loranthaceae and Viscaceae constitute 98% of mistletoe species. Among 1300 species of mistletoes reported 940 species are from Loranthaceae and 350 species are from Viscaceae (Watson and Dallwitz, 1992, Nickrent, 2001).They are wide spread in all five continents. They have been observed in tropical, temperate and arid zones. Many islands like Madagascar, Cameron, Hawaiian Islands, Fiji islands etc. have their own endemic species.

In India, mistletoes have been seen from sea level to 3500m in the Himalayas. Around 70 species have been reported by Pundir et al.(1997). They are a menace in forest, plantations and orchards.

Mistletoes have sacred mythical roles in several cultures (Kunwar et al. 2005). They are considered as keystone species of the ecosystem. They provide nectar and feed to insects and mammals. Medicinal properties of mistletoes have been utilised in the treatment of cancer, arthritis and other ailments.

In India, mistletoes of the genus, *Dendrophthoe* and *Viscum* are common while in East and West Africa genus *Tapinanthus* (Loranthaceae) is more common. In USA Dwarf mistletoes or *Arceuthobium* (Loranthaceae) is the common species. All these species affect high value tree crops such as teak, rose wood, eucalyptus, casuarina, neem etc. and fruit trees such as mango, sapota, guava, gooseberry etc. They are seen in coffee, tea, cocoa and rubber plantations.

Major difference between Lorantahceae and Viscaceae

Plant characteristics	Loranthaceae	Viscaceae
Calyculus	Present	Absent
Pollen grains	Triradiate	Spherical
Embryosac	Polygonous	Allium type
Embrygeny	Biseriate	Many tiered
Embryo suspensor	Present	Absent
Fruit	Viscid layer present outside vascular system	Viscid layer present inside vascular system

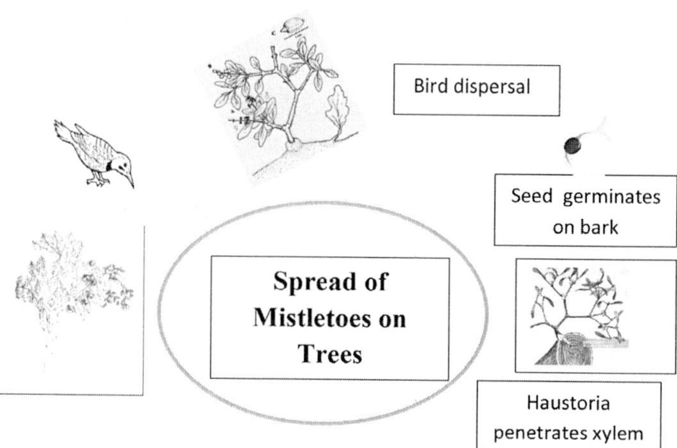

Fig 4: Life cycle of *Dendrophthoe falcata*

Birds are the main dispersal agents. Tickel's flower pecker is reported to facilitate seed dispersal. The haustorium connects to the xylem of the host plant. Infection affects the host foliage, phenology, respiration and productivity. It also affects wood quality and in the long run leads to death of the tree.

Common Genera and species of Loranthaceae and Viscaceae Seen in S. India

Loranthaceae	Viscaceae
Dendrophthoe around 70 species *D. falcata*, (L.f) *Etting, D. trgona*, (Wight&Arn) Danser ex Sant, *D. neelgherrensis,* W&A	*V. monoicum*, Roxburgh ex Candolle. *V. orientale*, Willd. *V heyneanum*, Willd *V. capitellatum*, Smith in Rees *V. ramosissimum*, Wall. *V. articulatum*, Burm. *V. angulatum*. Heyne ex DC.
Macrosolen-*M. parasiticus* *M. capitellatus*, (Wight&Arn.) Danser	
Helicanthus- *H. elastica,* (Desr) Danser	
Helixanthera- *H. obtusatus,* Wall *H. wallichianus,* Schutes., *H. intermedia,* (Wight) Danser	
Scurrula-*S.parasitica,* Desv.:W&A	
Tolypanthus-*T.lageniferus,* Wt	
Taxillus- *T.tomentosus,* Heyne *T.cuneatus* (Heyne ex Roth) Danser	

Control

Mistletoes are seen on high branches of tall trees. Hence it is difficult to devise suitable control measures for the parasite without harming the host plant. Pruning of infected shoots and branches is the commonly adopted method.

Spraying ethephon (25ml/L) on the parasite along with a nonpolar adjuvant like Organosilicone at (0.5ml/L) is found to cause defoliation and drying of the parasite without affecting the host plant. Spray must thoroughly wet the mistletoe foliage, Ideal time for treatment is just after the rains before the tree starts to bloom (Gargi, 2021).

Other popular method is swabbing the infected area with a paste of 2,4- D For this, dissolve I gm 2,4-D in 20ml water,soak a sponge or cotton in this solution. This is tied to the point of infection after scrapping the bark of the parasite. Note that 2,4-D is tied to the bark of parasite and not the host. Phytotoxicity symptoms like bark splitting can be seen if the chemical is applied on the host.

In dwarf mistletoes biocontrol agents such as fungul pathogen *Colletotrichum gloeosporiodes, Cylindrocarpon gillii, Caliciopsis arceuthobii* and *Neonectria neomacrospora* were found to be effective in destroying the aerial shoots of the parasite (Knutson and Hutchins, 1979; Kope and Shamoun, 2000; Muir and Moody, 2002)

References

Abu-Irmaileh, B. E. 1981. Response of hemp broomrape (Orobanche ramose) infestation to some nitrogenous compounds. Weed Science 29 ;8-10

Abu-Irmaileh, B. E. 1994. Nitrogen reduces branched broorape (*Orobanche ramose*) seed germination. Weed Science 42; 57-60.

Chancellor, R.J., Parker, C. and Teferedegn, T. 1971. Stimulation of dormant weed seed germination by 2-chloroethl phosphonic acid. Pesticide Science 2: 35-37.

Chinnusamy. C, Rathika .S. and Prabhakaran N.K. 2014 Witch weed (*Striga* spp.) In Naidu V.S.G.R and Mishra J. S.(Eds.) 2014. Parasitic Weeds Biology and Mnagement,114p. Today & Tomorrow's Printers and Publishers

Chinnusamy C.2006 Management of Cuscuta chinensis in Vegetable greens, pp 93-94.Annual report. AICRP -Weed Control, TNAU, Coimbatore.

Garggi, G. 2021. Physiological and molecular studies on genera of Loranthaceae and their management. Ph.D.Thesis, Kerala Agricultural University, Thrissur. 130p.

Gaur R.D. 1999. Flora of the District Garhwal, North West Himalaya, Transmedia, Srinagar Garhwal, pp443-444

Jayasinghe, C., Wijesundara, D.S.A, Tennekoon, T.U. and Marambe, B. 2004. Cuscuta species in the lowlands of Sri Lanka, their host range and host-parasite association. Tropical Agricultural Research 16:223-241.

Johnson, A.W., Rosebery, G. and Parker, C. 1976. A novel approach to Striga and Orobanche control using synthetic germination stimulants. Weed Research, 16(4):223-227.

Kanampiu, F.K., Friesen, D.K., Ransom, J.K. and Gressel, J. 2001. Imazethapyr seed dressings for Striga control on acetolactate synthase target-ste resistant maize. Crop Protection, 20:885-895.

Knutson, D. M and Hutchin, A. S. 1979. Wallrothiella arceuthobit infecting Arceuthobium douglasii; culture and field inoculation. Mycologia.71:821-

Kope H H and Shamoun S F. 2000. Mycoflora association of western hemlock dwarf mistletoeplants and host swellings collected from southern Vancover Island, British Colombia.Canadian Plant Disease Survey 80:144-147

Kroschel, J. 2001. A Technical Manual for Parasitic Weed Research and Extension. Kluwer Academic Publishers, Dordrecht, The Netherlands, 256pp.

Kunwar, R.M., Adhikari, N., Devkota, M.P. 2005. Indigenous use of mistletoes in tropical and temperate region of Nepal, Banko Janakari 15:38-42

Mishra J.S. 2009. Biology and management of Cuscuta species. Indian Journal of Weed Science, 41(1&2): 1-11.

Muir, J.A. and Moody, B. M . 2002. Dwarf mistletoe surveys USDA Forest Service Ge. Tech. Rep.RMRS-GTR-98, 67-116.

Nickrent, D. L. 2001. Mistletoe Phylogenetics: current relationships gained from analysis of DNA sequences. In Proc.West.Int.For.Dis.Work Conf., ed. B Geils, R Mathiasen, USDA For.Serv., Kona, HI

Pundir, Y.P.S. 1994.The magnitude of parasitism and the damages caused by Scurrula pulverulenta (Wall) G. Don (loranthaceae) in the Doon Valley and adjacent areas. World Weeds 1: 65-74

Pundir, Y. P. S. 1997a. Leafy mistletoe-Taxillus vestitus loss assessment in Chakrata oak forests. World Weeds.4:1-8.

Punia, S.S. 2014. Biology and control of Orobanche. Indina Journal of Weed Science, 46(1):36-51.

Ramachandra Prasad, T.V., Chinnusamy, .C. and Girija, T.2016. Parasitic weeds and their management. In Yaduraju, N. T. Sharma, A. R. and Das, T. K. (Eds.).2016. Weed Science and Management.402p.Indian Society of Weed Science, Jabalpur and Indian Society of Agronomy, New Delhi

Ramachandra Prasad, T.V., Mishra, J.S. and Girija, T. 2015. Management of parasitic weeds. Souvenir – 25th Asian Pacific Weed Science Soceity Conference, Telnagana State Agricultural University, Hyderabad, India, October 16-18, 2015. p.16-21.

Saghir, A.P., Foy, C.L., Hameed, K.M., Drake, C.R. and Tolin, S.A. 1973. Studies on biology and control of Orobanche ramose L. In Proceedings of European Weed Research Council Symposium on Parasitic Weeds. Malta, pp. 106-116

Sandip S. Nikam, Santosh B. P. and Kanade, M. B. 2014. Study of Cuscuta reflexa Roxb. with reference to host diversity, anatomy and biochemistry. Central European Journal of Experimental Biology, 3 (2):6-12.

Saunders A.R. 1933. Studies on phanerogamic parasitism,with particularreference to S.lutea Lour.South African Department of Agricultural Science Bulletin NO.128

Sheoran, Parvender, Punia, S.S., Samunder Singh and Dhiraj Singh. 2014. Orobanche weed management in mustard: Opportunities, possibilities and limitations. Journal of Oilseed Brassica, 5(2):96-101.

Trivedi, G.C., Patel, R.B., Patel, B.D., Meisuriya, M.I. and Patel, V.J. 2000. Some problematic weeds and their management – A review. Agricultural Reviews, 21(4):238-243.

Watson, D.M. 2001. Mistletoes- a keystone resource in forest and woodlands worldwide. Annu. Rev. Ecol. Syst.32: 219-249.

Watson, L. and Dallwitz, M.J. 1992. The Families of Flowering Plants: Descriptions, Illustrations, Identification, and Information Retrieval. Version 27th Sept.2000.

15

Weed Management in Major Field Crops

Rice

Rice is the primary food source for half of the Asian population. Nearly 2,000 million in this region obtain 60 to 70 per cent of their calories from rice and its products. Considering the importance of the crop, United Nations General Assembly declared 2004 as the "International Year of Rice" (IYR). In India, the crop covers one-fourth of the total cropped area. About half of the Indian population use rice as their staple food.

Rice based cropping systems are not only important for food security but also for maintaing an ecological balance. Low land rice fields conserve water and prevent runoff .They preserve the sustainability of the ecosystem.

It is a highly adaptable crop which grows between 8° to 25° N latitude and from 2.2 m below sea level in Kuttanad of Alleppy district in Kerala to about 3050m above sea level in Jumla situated in Karnali province of the Himalayan mountainous region. Chumchaur of Jumla in Nepal is the place where rice cultivation is carried out at the highest elevation in the world. It is a tropical plant and requires high heat and high humidity for its successful growth and prefers an annual rainfall of 150 to 200cm. It can be grown in different soil types including silts, loams and gravels and can tolerate acidic as well as alkaline soils.

In India, the crop is cultivated under diverse conditions from Kerala to Kashmir and from Gujarat to the North Eastern states. Major constraint in rice cultivation is the need for plenty of labour. There are different seasons and different methods of cultivation followed in different states. The seasons may be classified as Ist crop season starting from May-June to September –October IInd crop season, starting from June-July to November-December and Summer crop starting from November-December to March April.

Method of cultivation also varies from region to region and also from season to season. These include direct seeded upland/dry sown rain fed system, direct seeded lowland rainfed rice, direct seeded rice under puddled condition and non-puddled condition, direct seeding using seed drills, transplanting under puddled condition and transplanting using machines. Broad casting in dry seeded areas as well as wet seeded areas. In all these systems, weed is the major bottleneck in rice cultivation.

Diversified rice growing situations of the country include upland, low land, saline soils, alkaline soils, acid sulphate soils, sandy soils, iron toxic soils etc. These varying situations in rice ecosystem have a profound influence on the weed flora. It is observed that weed species in rice ecosystem vary with soil, temperature latitude, altitude, rice culture, planting methods, water management, fertility levels and weed control practices.

Major weed flora in rice ecosystem

The weeds found in rice fields can be divided in to five general types based on their appearance and botanical features. They can be classified as grasses, sedges, broad leaved weeds, ferns and algae.

Upland rice ecosystem

Echinochloa colona, Sacciolepis interrupta, Digitaria ciliaris, Cleome viscosa

Dactyloctenium aegyptium, Eleusine indica, Cynodon dactylon, Sporobolus diander, Rottboellia exaltata, Paspalidium flavidum, Cyperus difformis, Cyperus rotundus, Commelina benghalensis, Portulaca oleracea, Cyperus iria

Wet land rice ecosystem

Echinochloa crusgalli, Echinochloa stagnina. Echinochloa glabrescens.

Isachne miliacea, Ischaemum rugosum. Hygrorhyza aristata. Leptochloa chinensis, Paspalum distichum, Cyperus iria, Cyperus compressus, Fimbrystylis miliacea, Schoenoplectus articulates, Schoenoplectus lateriflorus. Monochoria vaginalis. Eichornia crassipes. Bacopa monnieri, Eriocaulon quinquangulare. Limnocharis flava, Hydrilla verticillata, Nymphaea nouchali Pistia stratiotes, Ludwigia parviflora, Sphenoclea zeylanica, Lymnophila heterophylla Lymnophila repens, Centella asiatica. Sphaeranthus indicus. Wedelia calandulacea, Ammania baccifera, Rotala macrandra, Rotala indica, Alternanthera sessilis, Aeschynomene indica, Ceratopteris thalictroides.

Weedy rice or *Oryza sativa* f. *spontanea* is considered as an objectionable weed in rice. It is a major problem in direct sown rice.

Critical period of crop weed competition in rice ranges from 20-45 days after transplanting in transplanted rice and 15-60 days after sowing in wet seeded rice. In upland rice first 30 days are considered as critical. Weed control during initial stages is important.

Methods of Weeds Management

Using competitive cultivars.

Tall cultivars with high tillering and vegetative growth are reported to supress weed flora in rice ecosystems. Using high yielding competitive varieties along with other management practices is seen to reduce weed infestation in rice fields. Vandana, Kallinga- III, RR -151-3, Chennellu and Karanellu are some such varieties.

Water management

Maintaining the field in flooded condition after proper crop establishment is the best cultural operation to reduce weeds in wet land rice ecosystem. After herbicide application, maintaining the field in flooded condition would help in the decay of weeds.

Stale seedbed technique

This technique is most suitable for direct sown rice. Pre monsoon showers or even irrigating the field after land preparation would activate weed germination. These seedlings can be killed using non selective herbicides or shallow cultivation. This will deplete weed seeds from the soil seed bank.

Manual and Mechanical weeding

Hand hoes, blade hoes and wheel hoes are used in direct seeded rice. In transplanted crop, rotary weeder, cono- weeder, peg type dryland weeders are effective. However these equipments cannot be used within rows. Therefore mechanical weeding followed by hand weeding is recommended.

Chemical weed control

In the present scenario weeding is a labour intensive and costly operation. Scarcity of labour force and its financial implication have resulted in more and more farmers adopting chemical weed control measures. Several herbicides offer effective control of weeds. Judicious use of the chemicals in proper dose and at the appropriate time will help the farmer to control rice weeds effectively. Some of the chemicals commonly used in rice is given in table.

Table 1: Pre-emergence Herbicides Recommended for Rice

Formulation	**Dosage Kg ai.ha^{-1}**	**Time of application DAS/DAT**
Butachlor 50 EC	1.25	0 – 6
Pretilachlor 50 EC	0.75	0 – 6
Pretilachlor + Bensulfuron methyl (0.6 + 0.06)	0.6 +0.06	0 – 6
Pyrazosulfuron ethyl 10 WP	0.02 – 0.03	0 – 6

Table 2: Post-emergence Herbicide Recommended for Rice

Formulation	**Dosage Kg ai. ha^{-1}**	**Time of application DAS/DAT**
2,4-D 80 WP	0.80	20 – 25
Chlorimuron ethyl + Metsulfuron methyl 20WP	0.004	20 – 25
Carfentrazone - ethyl 40 DF	0.02	20 – 25
Cyhalofop butyl 10EC	75-80g ai.ha	15 – 18
Fenoxaprop-p-ethyl 6.95 SC, 9 EC	0.06	20-25 Weeds 3-5 leaf stage
Azimsulfuron 50 DF	17.5 g ai. ha^{-1}	15 – 20
Penoxsulam	0.025	15 – 20
Bispyribac Na 10 EC	0.025	15 – 20
Ethoxy sulfuron 15 WDG	0.015	15 – 20
Oxadiargil 80 WP, 6 EC	0.10	3-5 DAT
Florpyrauxifen benzyl	60g ai .ha^{-1}	12-18 DAS
Florpyrauxifen benzyl+Cyhalofop butyl EC	180g. ha^{-1}	12-18 DAS
Triafamone 20%+ethoxy sulfuron 10% WG	60-65.5 g.ai ha^{-1}	15 DAT
Butachlor1%w/v +penoxsulam +40%w/v EC	225g.ha^{-1}	10 DAT
Cyhalofop 5.1% + Penoxsulam1.02% EC	135gai.ha^{-1}	12-18 DAS

Integrated weed management

Integrating available techniques judiciously as per need and the situation is found to be the most successful technique for managing weeds in rice. In wet land cultivation puddling helps to kill weeds and retain water. When seedlings are transplanted into standing water it helps in getting a good stand density, flooded condition suppresses weeds. In such situation a single hand weeding before top dressing would be sufficient to get good weed control. Management of wild rice is given as case study in integrated weed management

Weed management in Nursery

Keeping the nursery weed free is important to prevent mimicry weeds getting planted along with rice. Hence solarisation or stale seed bed technique can be adopted to reduce contamination from nursery. Using weed free clean seeds for nursery preparation is also important.

Wheat

Wheat is one of the oldest and most prominent cereal crops in the world. In India, it is the second important food grain, next to rice. It is grown in most of the north Indian states. Introduction of short statured (Semi-dwarf) high yielding Mexican varieties had led to a dominance of grass weeds in wheat field during the post green revolution. To realise the full potential of such varieties, increased irrigation and fertiliser application is also necessary. In India, around 10.5 M ha area is under Rice –Wheat cropping system. These areas are in Punjab, Utter Pradesh, Bihar, Gujarat and Maharashtra. They play a major role in sustaining our food security. It was estimated that the average yield loss due to weeds in the different growing zones ranged from 20 to 30% depending on weed density and management practices (Chhokar et al.,2012)

Weed flora

Chenopodium album, Fumaria parviflora, Anagallis arvensis Argemone mexicana, Avena fatua, Asphodelus tenuifolius, Cynodon dactylon, Phalaris minor, Alopecurus myosuroides, Rumex dentatus, Cirsium arvense, Lolium multiflorum, Poa annua, Sinapis arvensis, Galium tricornutum, Ranunculus arvensis.

Critical period of weed competition

Critical period is influenced by the weed community (weed species, number of weeds, emergence date), cropping practices (cultivar, row spacing, population, tillage) and environment. In wheat a weed free period upto 45 days is recommended.

Methods of Weed Management

Cultural method

Use weed free clean seeds for sowing. This includes cleaning farm implements like seed drills used in farming operations. Type and intensity of tillage operations has a profound effect on the weed flora. Minimum tillage leads to seed build up in the upper soil layers while conventional tillage distributes the weed seeds in different layers. Zero tillage is found to reduce *P. minor* but continuous use of zero tillage leads to shift in weed flora and to *Rumex dentatus* and *Malva parviflora* (Chhokar et al., 2007).Crop rotation using sunflower, berseem, or sugarcane is found to reduce the population of *P.minor*. Close row planting, increasing seed rate and bidirectional sowing smother weeds. Adequate and balanced fertilizer increases the competitive ability of wheat crop to weed.

Mechanical method

Weeding at 20 and 40 days after sowing is recommended in wheat. This can be done using hand hoe or manually. Mechanical control of weeds during the vegetative growth period of wheat is very limited so chemical method is commonly used.

Chemical method

Herbicides Recommended for Wheat Crop

Herbicide	Dose ai. g.ha^{-1}	Time of application DAS
Dicamba	300-500	30-35
MCPA	500-1000	30-35
2,4-D amine	600	30-35
Pendimethalin 30% EC	750-1500	0-3
Sulfosulfuron	25	30-35
Carfentrazone	20	25-30
Metsulfuron+idosulfuron	12+2.4	30-35
Clodinafop15%+metsulfuron1%	160	35
Metsulfuron+carfentrazone	20+4	30-35
Isoproturon+metsulfuron	1000+4	30-35
Isoproturon+2,4-D	1000+400	30-35

Millets (Sorghum, Maize, Pearl millet)

Millets are nutritionally healthy cereals which can be grown in resource poor agro-climatic regions. They are termed as' nutri-cereals' The United Nation General Assembly has declared 2023 as the International year of millets to raise the awareness of the nutritional and health benefits of millets and to sustain its production. Weeds are a major deterrent in sustaining productivity of the crop. Average yield loss due to weeds is estimated to be 15-83% in sorghum, 16-94% in pearlmillet and 55 to 61% in finger millet (Mishra et al 2016)

Major millets are Sorghum (*Sorghum bicolor*) and Pearl Millet

(*Pennisetum glaucum*) and Minor millets include Finger millet (*Eleusine coracana*), Kodo millet (*Paspalam scorbiculatum*), Proso millet (*Panicum miliaceum*), Barnyard millet (*Echinochloa frumentaceae*), Foxtail millet (*Setaria italica*)and Little millet(*Panicum sumatrense*).

Weed flora in millets depends on the season, agro ecological region and management practices followed. They do not compete well with weeds as the seedlings are small and their growth is very slow during the initial 20 to 25 days, which is also considered as the critical period of weed growth.

Major weeds

Cynodon dactylon, Brachiaria ramosa, Digitaria sanguinalis, Chloris barbata, Dactyloctenium aegyptium, Dinebra retroflexa, Eleusine indica, Echinochloa colona, Sorghum halepense, Setaria glauca, Panicum viridis, Panicum repens, Paspalam paspaloides. Convolvulus arvensis, Amaranthus viridis, Cyperus rotundus, Ageratum conyzoides, Celosia argentea, Cleome viscosa,

Trianthema portulacastrum Achyranthes aspera, Commelina benghalensis, Tridax procumbens, Striga spp., *Euphorbia heterophylla, Euphorbia hirta, Eclipta alba, Xanthium strumarium*

Methods of Weed Management

Cultural method

Growing pulses such as mung bean,groundnut, cowpea and soybean as intercrops is found to supress weeds. Narrow row spacing, higher seed rate, early nitrogen application and placement of fertiliser near the crop increases plant vigour and helps to smother weed population.Manual and mechanical weeding is also recommended

Chemical method

It is observed that pre emergence application of herbicides is more effective in millets. Only 2,4-D and paraquat are recommended as post emergence sprays. Most commonly recommended herbicides are given below.

Table 4: Herbicides recommended for Sorghum and Maize

Chemical	Dose (kg.ai.ha^{-1})	Time of application
Atrazin	0.75-1.0	Pre-emergence
Simazine	1.5-2.0	Pre-emergence
Metolachlor	1.0-1.5	Pre-emergence
Atrazine+ Pendimethalin	0.75+0.75	Pre-emergence
MCPA	3.5	Pre-emergence
Atrazine+Metolachlor	0.75+0.50	Pre-emergence
2,4-D	0.50-0.75	Post emergence
Paraquat	0.2-0.5	Post emergence directed spray

Reproduced from Mishra (2015)

Table 5: Herbicides recommended for Millets

Chemical	Dose (kg.ai.ha^{-1})	Time of spray
Pearl millet		
Atrazine	0.50	Pre-emergence
Pendimethalin	0.50-0.75	Pre-emergence
Oxadiazin	1.0	Pre-emergence
2,4-D	0.50-0.75	Post emergence
Finger millet		
Isoproturon	0.50-0.75	Pre-emergence
Oxadiazin	1.0	Pre-emergence
Kodo millet		
Isoproturon	0.50	Pre-emergence

Reproduced from Mishra (2015)

Pulses

A variety of pulses are grown in different climatic zones of the country. Pegionpea, greengram, blackgram and cowpea are grown during both rainy season and summer while chickpea, fieldpea, lentil, lathyrus and rajmash are grown in winter both in irrigated and rainfed ecosystems. They are also grown as intercrops in many situations for effective utilisation of land and water resources. Weeds are a major constrain in realising the potential yield of the crop. It has been estimated that weeds contribute to 31% loss in pigenpea, 110% loss in urdbean and 60% loss in mung bean.(Ali and Lal.1989)

Major weeds

Alternanthera sessilis, Cleome viscosa, Aerva lanata, Celosia argentea,

Trianthema portulacastrum, Achyranthes aspera,, Commelina benghalensis, Ageratum conyzoides, Corchorus acutangulus, Ashphodelus tenuifolius are common weeds in chickpea under rainfed condition. *Convolvulus arvensis* is reported from field pea, chickpea and lentil. *Philaris minor* from irrigated crop in North India and *Cuscuta* spp. from black gram and green gram fields of Telangana and Tamilnadu.

Table 6: Critical Period of Crop Weed Competition and Yield Reduction in Pulses

Crop	Critical period of growth	Reduction in Yield %
Pigenpea	15-60	24-40
Green gram	15-30	30-50
Black gram	15-30	30-50
Chickpea	30-60	15-35
Lentil	30-60	20-30
Pea	30-45	20-30

Reproduced from DWSR, 2018

Methods of Weed Management

Cultural method

In wide spaced pulse crops like pigeon pea, intercropping with sorghum, maize, green gram, black gram and groundnut is a common practice. Hand weeding and hoeing are common practices in pulses.

Table 7: Crop and Recommended Time of Weeding in Pulses

Crop	Recommended Time of weeding
Pigeon pea	2 weeding one at 25-30DAS and another at 45-50DAS
Mung bean	25 and 45 DAS
Winter pulses like chickpea, lentil, French bean and faba bean	30 and 60DAS

Chemical method

Table 8: Herbicides Recommended for Pulses

Herbicide	Dose (kg.ai.ha^{-1})	Time of application
Pendimethalin	0.75-1.20	Pre-emergence
Imazethapyr	0.075-0.150	Pre -emergence
Oxyfluorfen	0.1-0.15	Pre-emergence
Diclosulam	0.037-Soybean 0.01- Mungbean	Pre-emergence
Quizalofop-p-ethyl	0.0375 -0.050	Post emergence
Haloxyfop-p-methyl	0.075-125	Post emergence
Imazethapyr	0.05-0.075	Post emergence

Reproduced from Sinchana and Raj (2020)

Sugarcane

Sugarcane is an ancient crop introduced into India by Austronesian traders around 1200 to1000 BC. Sugarcane is cultivated in most states of India. Nearly 4.8 million ha. Area is under sugarcane cultivation in India(Co-operative sugar, 2022). It is a perennial crop which remains in the same field for 3 to 4 years. The crop duration is 12 to 18 months. Initial growth is very slow and so it is infested with a variety of weeds. Nearly 150 weed species have been reported from sugarcane fields. However weed flora differs with the agro climatic condition of the state. In sugar cane critical period of weed control is 90 to 120 days after sowing (Shali et al.,1994).

Major Weed flora

Echinochloa colona, E. crusgalli, Dactyloctenium aegyptium, Celosia argentia, Amaranthus viridis, Cyperus rotundus, Cynodon dactylon, Chenopodium album, Angallis arvensis, Fumaria parviflora, Sorghum halepense, Lathyrus sativa, Vicia spp.

Methods of Weed Management

Cultural and mechanical

Sugarcane is a widely spaced crop. Hoeing is an important operation in sugarcane cultivation. It not only removes weeds but also activates tillering, destroys insects and enhances aeration in the soil. As it is a laborious and costly operation some growers grow a second crop in between the sugarcane crop as a mixed crop to reduce cost of production. Crop rotation with fodder crops like sorghum will help to eliminate associated weeds of sugarcane. Intercropping mustard, potato or wheat in autumn, and black gram, green gram or cowpea in spring planted sugarcane has been found to be effective in suppressing weeds of sugarcane crop. Trash mulching is adopted to supress weeds, conserve moisture and provide organic matter to soil. Sugarcane leaves are used as thrash soon after emergence of the cane, to cover the soil between the cane rows.

Chemical method

Sugarcane is infested with a large variety of weeds. Application of a single herbicide is not effective. It is seen that pre emergence application of any one of the herbicides given below followed by post emergence sprays gives a better control of weeds.

Table 9: Herbicides Recommended for Sugarcane crop

Herbicide	**Dose (kg.ai.ha^{-1})**	**Time of application**
Simazine	1.5-2	Pre-emergence
Atrazine	1.5-2	Pre-emergence At 3DAS
Metribuzine	1.0	Pre-emergence
Diuron	2.5-3	Pre-emergence
Ametryn	2.0	Pre-emergence
2,4-D	2.5-3	Post emergence
Halosulfuron methyl 6% + Metribuzin 50% WG	1.25	Pre and Post emergence

Weed management in sugarcane ratoon

Ratoon crop of sugarcane is highly infested with weeds. Normally yield of ratoon crop is 20 to 30% lesser than the plant crop, mainly because weeds from the previous crop have an upper hand in utilizing various inputs and natural resources. Three hoeings at 30, 60 and 90 days after harvest is reported to give maximum ratoon yield. Hoeing can also be replaced by atrazine or 2,4-D sprays immediately after harvest of the plant crop depending on the weed flora.

Integrated weed management

Integrated weed management in sugarcane involves maintaining the fallow fields weed free by thorough field preparation before planting. This includes pre- emergence spray of atrazine or any broad herbicide to reduce weed growth, ensuring good and even germination of the planted canes, avoiding gaps and trash mulching to reduce weed emergence. Selective manual weeding is done before fertiliser application to enhance cane growth followed by post emergence spray of herbicides to control broad leaved weeds or directed sprays of contact herbicides for reducing weed competition.

Cotton

India has the largest cotton area in the world. It is a perennial tropical crop which prefers warm temperate climate. The crop is sensitive to weed competition during the early stages of growth.

Weed Flora

Trianthema portulacastrum, Celosia argentea, Echinochloa colona, Digera arvensis Brachiaria ramosa, Digitaria sanguinallis, Cenchrus catharticus, Chloris barbata Eleusine aegypticum, Phyllanthus niruri, Euphorbia spp., *Tridax procumbens, Vernonia cinerea, Corchorus acutangulus, Abutilon indicum, Achyranthus aspera, Aristolochia bractata, Alysicarpus rugosus, Desmodium diffusum.*

Cultural methods

Cotton is a long duration crop that remains in the field for 160 to 300 days. Moreover cotton has wide row spacing (60 to120cm) depending on the variety. Hence intercropping of short duration quick growing crops like cowpea, green gram, black gram, soybean and groundnut can be done to supress weeds.

In early planted summer crop, growth of weeds is meagre due to lack of water. Weeds that emerge with the summer rains or irrigation can be easily controlled by weeding. When proper monsoon arrives the cotton crop will be fully established and the canopy will be dense enough to supress weeds.

Chemical method

Table 10: Herbicides Recommended for Cotton

Herbicide	Dose (g.ai.ha^{-1})	Time of application
Fluchloralin	1000	Pre-plant incorporation
Trifluralin	500-1000	Pre-plant incorporation
Butachlor	1000-1250	3-4DAS
Pendimethalin	1000-1500	3-4DAS
Diuron	500-750	3-4DAS
Oxadiazon	500-700	3-4DAS
Fluazifop-butyl	500	Post emergence
Glufosinate	450	Post emergence
Glyphosate	1000	Post emergence
Pyrithiobac-Na	60-70	Post emergence

Oil Seeds

Sesamum

Sesamum is an ancient crop of India with an area of 16.23lakh ha and 6.58Lakh tons productivity in 2019-20 as per data given by the Directorate of economics and statistics. The area under sesamum cultivation shows a decline. Weeds are a major biotic constraint in the production of sesamum. Slow growth of sesame seedlings during first four weeks, makes it a poor competitor in the field (Nazir, 1994; Bennett et al., 2003). Weed competition alone contributes to 81 per cent yield loss (Shaalan et al., 2014). Slow initial growth of sesame and intermittent rains are conducive environment for weed growth. Hence weed control at early crop periods mainly at the critical period of crop weed competition (20-30 DAS) is essential to maintain yield in the crop (Weaver et al., 1992).

Major weed flora

Melochia corchorifolia, Digitaria ciliari, Echinochloa colona, Cleome viscosa. Amaranthus viridis, Celosia argentea, Chloris barbata, Corchorus olitorius, Cynodon dactylon, Cyperus rotundus, Cyperus iria

Methods of Weed Management

Agronomic methods

Improving germination % of seeds and vigour of seedlings was found to be effective in case of broadleaved weeds such as *Melochia corchorifolia.* Seed priming with 100ppm MnSO4 or 0.3% Borax was found to significantly reduced the weed count as compared to control both at 10 DAS and at harvest.

Weed control efficiency was found to be maximum for $MnSO_4$ followed by priming with borax indicates that these priming treatments could give 30-38 per cent weed control (Sreepriya and Girija, 2018)

Two hand weedings at 15 and 35 DAS is essential

Table 11: Herbicides Recommended for Sesamum crop

Chemical	Dose g ai.ha^{-1}	Time of application
Diuron	400-600	Pre emergence
Pendimethalin	1000	Pre emergence
Oxyfluorfen	100	Pre and post emergence

Integrated weed management in Sesamum

A well-drained loose loamy or sandy soil is ideal for sesamum cultivation which is also preferred by weeds like *Melochia corchorifolia* and *Cleome viscosa*. To reduce weed infestation integrated approach gives better results. After giving a fine tilth to the soil, weeds from the soil seed bank can be encouraged to germinate by irrigating the field.These weeds can be removed by application of a broad spectrum herbicide like glyphosate. After application of stale seed bed technique, sesamum seeds can be sown. Priming the sesamum seeds with $MnSO_4$ will improve germination rate and also vigour of the germinated seedlings. Use of herbicides like oxyfluorfen immediately after germination of sesamum seeds would help to reduce germination of weed seeds. Selective manual weeding just before top dressing is also recommended.

Groundnut

Groundnut is an important oil seed crop of India. Cultivation is scattered in 12 states in diverse climatic conditions. It is grown throughout the year. Weed flora in the crop varies with the agro-climatic conditions of the state.

Critical period of weed competition is reported to be 4 to 8 weeks (Bhan et al 1976, Yadav et al 1982)

Weed flora

Cynodon dactylon, Trianthema portulacastrum, Celosia argentea, Cyperus rotundus, Achyranthus aspera, Brachiaria cruciformis, Dactyloctenium aegyptium, Sida acuta, Digitaria sanguinalis, Dinebra retroflexa, Eleusine indica, Echinochloa colonum, Setaria spp, *Panicum repens, Cleome viscosa, Commelina benghalensis, Brassica* spp. *Cyanotis cucullata, Mollugo disticha, Mollugo pentaphylla, Parthenium hysterophorus, Euphorbia geniculata, Euphorbia hirta, Phyllanthus amara.*

Methods of Weed Management

To improve productivity of groundnut the field should be kept weed free upto 60 days. First weeding is recommended by 20-25 DAS followed by second weeding at 45 DAS. Manually operated instruments like star weeder and hand hoe are reported to improve efficiency of weeding. Intercultural operations like harrowing also loosen the soil and improve pegging in groundnut. Duck foot harrow and power operated tillers are also used for intercultural operations in groundnut. These operations also help in soil moisture conservation and infiltration to the root zone. According to Shetty and Rao (1981), increase in density of intercrops like sorghum or pigeon pea can reduce weed growth and can replace one hand weeding. Dense planting and crop rotation with Maize or sorghum and use of polythene mulches have also been recommended for reducing weed population in groundnut (Devi Dayal et al.,1994)

Table 12: Herbicides Recommended for Groundnut

Herbicide	Dose (kg a.i./ha)	Time of application
Pendimethalin	1.0-1.5	Pre-emergence
Butachlor	0.75-1.0	Pre-emergence
Anilofos	0.5	Pre-emergence
Oxyflorefen	0.25	Pre-emergence
Metalochlor	1.0	Pre-emergence
Imazethapyr	.075	Post –emergence2 0DAS
Quizalofop-ethyl	0.05	Post –emergence 20DAS
Fluazifop-p-butyl	0.25	Post –emergence 25-30 DAS

(Source: Devi Dayal, 2004; AICRPG, 2009)

References

AICRP, 2009. Biennial Progress Report (2006-08). All India Coordinated Research Project on Management of Salt-affected Soils and Use of Saline Water in Agriculture, CSSRI, Karnal, pp. 212. (6) https://www.researchgate.net/publication/231538151 Performance of groundnut Arachis hypogaea Cowpea Vigna unguiculata cropping system under various salinity levels in black clay soils of Saurashtra Gujarat

Ali, M. and S. Lal, 1989. Priority inputs in pulse production. Ferfil. News 34 (11): 17-21

Anusha S., Tayade, A. S. and Geetha, P. 2018. Weed management for sustainable sugarcane production. In: Best management practices for sustaining sugarcane productivity. ICAR-Sugarcane Breeding Institute, Coimbatore. pp: 151-158

Bennett, M., Katherine and Conde, B. 2003. Sesame recommendations for the Northern Territory. Agnote, 657: 1-4

Bhan, V. M., Negi, P. S. and Singh, G. R. 1976. Effect of metoxuron on the control of phalaris minor in wheat and its residual effect on subsequent crops. Indian J. Weed Sci., 88888: 70-73

Channappagoudar, B.B. and Biradar, N.R. 2007. Physiological approaches for weed management in soybean and red gram (4:2) intercropping system. Karnataka J. Agric. Sci.20: 241-244.

Chauhan, B.S. and Johnson, D.E. 2010. The role of seed ecology in improving weed management strategies in the tropics.Adv. Agron. 105: 221-262.

Chhokar, R. S., Sharma, R. K., & Sharma, I. 2012. Weed management strategies in wheat-A review. Journal of Wheat Research, 4(2), 1-21.

Co-operative Sugar .2022.Vol.53, August Updated August 2022; Vol. 53, No.12. (Published in Cooperative Sugar)

Devi Dayal, Naik, P. R., Dongre, B. N. and Reddy, P. S. (1994). Effect of row pattern and weed control method on yield and economics of rainfed groundnut. Indian J. Agric. Sci.., 446-449.

Devi Dayal .2004. Weed management in groundnut. In; Groundnut Research in India by Basu, M. S. and Singh, N.B.PP. 248-259.

Directorate of Economics and Statistics, Department of Agriculture, Cooperation and Farmers Welfare, Ministry of Agriculture and Farmers Welfare, GOI. http://eands.dacnet.nic.in

DWSR (Directorate of Weed Science Research). 2018. Weed Management in Improving Agricultural Production. ICAR-Directorate of Weed Science Research, Jabalpur, Madhya Pradesh, 57

Jayan. 2018. India loses farm produce worth 11b$ to weeds every year [on-line]. Available: http://www.thehindubusinessline.com[15 Jan. 2018]

Mishra, J.S., Rao A.N., Singh, V.P., and Rakesh Kumar.2016.Weed management in major crops. In Yaduraju, N. T. Sharma, A. R. and Das, T. K. (Eds.).2016. Weed Science and Management.402p. Indian Society of Weed Science, Jabalpur and Indian Society of Agronomy, New Delhi

Mishra, J.S. 1997. Critical period of weed competition and lossesdue to weeds in major field crops. Farmers ant Parliament XXXIII 6: 19-20.

Mishra, J.S 2015. Weed management in millets: Retrospect and prospects. Indian Journal of Weed Science 47(3): 246–253,

Nazir, M.S. 1994. Crop Production. National Book Foundation, Islamabad, Pakistan, pp. 359.

Pala, Fırat and Mennan, Hüsrev. 2021. Common Weeds in Wheat Fields. https://wikifarmer.com/weed-management-in-wheat-farming/

Rao, A. N. and Shetty, S. V. R. (1981). Investigation on weed suppression ability of smoother cropping system in relation to canopy development and light interception. In: Proc. Eighth Asian Pacific Weed Science Society Conference, PP. 357-364.

Shaalan, A.M., Abou-Zied, A.K. and Elnass, M.K. 2014. Productivity of sesame as influenced by weeds competition and determination of critical period of weed control. Alex. J. Agric. Res., 59: 179-87.

Shali M, Afghan S, Shah M, Mahmood.1994.Screening of herbicides for weeds in sugarcane at post-emergence stage. Pak Sugar J. ;8(1):9-12.

Shetty, S.V.R. and Rao, A.N. (1981). Weed management studies in sorghum/pigeon pea and pearl millet/groundnut intercropping systems- some observations. In: Proc. Intl. Workshop on Intercropping, Hyderabad, India, January 10-13, PP. 238-248

Sinchana, J.K, and Raj, S. K. 2020. Weed Management in Pulses: A Review .Legume Research-An International Journal, Volume Issue 10.18805/LR-4375:1-8

Sreepriya, S. and Girija, T. 2018. Seed priming for improving the weed competitiveness in sesame. Journal of Crop and Weed, 14(2): 40-45 (2018)

Sundara, B.1998. Sugarcane cultivation. Vikas Publishing House pvt Ltd.

Sushilkuamr. 2009. Biological control of Parthenium in India: status and prospects. Indian J. Weed Sci. 41 (1 and 2): 1-18.

Tayade, A.S., Geetha, P., Dhanapal, R., Hari, K. 2016. Effect of in-situ trash management onsugarcane under wide row planting system.Journal of Sugarcane Research 6(1): 35-41.

Weaver, S.E., Kropf, M.J., and Groeneveld, R.M.W. 1992. Use of ecophysiological models for cropweed interference: The critical period of weed interference. Weed Sci., 40: 302–307.

Yadav. S. K., Singh, S. P. and Bhan, V. M. (1982). Effect of weed removal on pod yield of groundnut. Abs. Annual Conference of Indian Society of Weed Science (undated) PP. 15.

Yaduraju, N.T. and Ahuja, K.N. 1990. Weed control through soil solarization. Indian J. Agron. 35(4): 440–442.

16

Weed Management in Horticultural Crops

Tea Gardens

Indian tea has a global reach being the second largest tea producer in the world.

Indian tea is one of the finest in the world. Darjeeling tea is the first product to get a geographical indicator tag in India. 83 % of the tea produced in India is from Assam and West Bengal and 17% from the south Indian states of Kerala, Karnataka and Tamilnadu. Weeds are a major problem in tea gardens.

Weed flora

Drymaria cordata, Bidens pilosa,, Cynodon dactylon, Panicum repens, Celosia argentea, Ageratum conyzoides, Centella asiatica, Borreria hispida, Oxalis latifolia, Mikania micrantha, Paspalam conjugatum, Drymaria diandra Eupatorium adenophorum, Imperata cylindrica, Mitracarpus verticiliatus Cyperus rotundus, Setaria palmifolia, Emilia sonchifolia, Artemisia vulgaris Conyza ambigua, Syndnedrella nodiflora, Crassocephalum crepidioides.

Methods of Weed Management

Manual weeding or weeding with hoe is the age old custom of weed removal adopted in tea gardens. These methods are eco- friendly and add to improving the nutrient content of soil. However they are not cost effective and scarcity of labour has compelled the farmer to look for cheaper alternatives.

Soil flipping and mulching are also adopted by farmers. Soil flipping overturns the weeds and creates a hostile environment for them to grow. The major disadvantage is that it encourages soil erosion and involves human labour.

Chemical weed control is the most cost effective technique. Major problem is that chemicals used for weed control may also be absorbed by the tea plants and may be passed on to the consumers. Simplicity of application and effectiveness of the chemicals has contributed to the wide adoption of chemical herbicides for weed control in tea estates by the farmers.

Chemical Method

Table 1: Herbicides Recommended for Tea Gardens

Herbicide	Dose Kg.ai. h^{-1}	Time of application
Dalapon	1-1.5%	Post -emergence
Simazine	2.25-3.0	Pre-emergence
Diuron	2.62-3.40	Inter row spraying
Paraquat	0.1% or 0.25-0.4	Directed and spot sprays
Glyphosate	0.6-1.0	Directed and spot sprays
Glufosinate ammonium	0.3-0.5	Directed and spot sprays
Carfentrazone ethyl0.43%+Glyphosate30.82%EW	12.90+924.60	Directed and spot sprays
Oxyfluorfen	0.12-0.25	Both pre and post-emergence

Before Planting

Tea is normally planted in high ranges where the terrain is uneven and covered with perennial grasses and forest trees. The land is cleared by cheeling or sickling. After clearing the land perennial grasses are controlled by treating with dalapon or glyphosate depending on the species, 6 to 8 weeks before planting.

Nursery management

Weed control in clonal nursery is done by pre plant application of simazine applied 2 to 3 weeks before planting. This maintains weed free situation for 4-6 months. Later when the plants are established and put 2 flushes of growth 6 months after planting any post - emergence herbicide may be applied as directed spray to remove the existing weeds.

Management in young plantations

Till the plants are 3 years old they are called young plants. Weeds grow intensively during this period. In such plantations simazine is applied as pre-emergence on clean soil just before the early rains between March- April to prevent the growth of annual weeds. Post emergence targeted sprays can be given using any of the post emergence chemicals recommended such as paraquat, glyphosate or glufosinate ammonium.

Established plantations

In established plantations surface roots occupy the surface area so intercultural operations using mechanical means is difficult. Root injury leads to soil born infections. Simazine, diuron and oxyfluorfen are pre-emergence herbicides used in established tea gardens. Normally herbicides are applied before rains in March- April and late December- January. Other post emergence herbicides are applied as per requirement.

Coffee Plantations

Indian coffee grows mainly in four South Indian states under monsoon rainfall conditions. It was first introduced in Chikmagalur hills of Karnataka by a sufi saint, 'Baba Budan' from Yemen in 1670. Karnataka accounts for 53% of coffee production followed by Kerala, 28% and Tamilnadu 11%. Most of the produce is exported. The two well-known coffee species are 'Arabica' and 'Robusta'.

Weed Flora

Digitaria marginata, Paspalum conjugatum, Ageratum conyzoides, Borreria hispida, Euphorbia hirta, Commelina benghalensis, Eleusine indica. Parthenium hysterophorus, Tagetes minuta, Solanum nigrum. Bidens pilosa, Amaranthus sp., *Galinsoga parviflora.*

Methods of Weed Management

Digging and forking is done in coffee plantations. during the months of September- November and February- March. Dalapon is used to control grass weeds and 2, 4-D amine salt to control broad leaved weeds,

Young coffee fields

Application of atrazine, simazine or diuron a few days before planting seedlings with minimum soil disturbance is found to give good results.

Established plantations

Pre- emergence spray of simazine, atrazine or diuron before the onset of rains gives good control. After monsoon season, targeted sprays of contact herbicides like paraquat is recommended.

Table 2: Herbicides Recommended for Coffee Plantations

Herbicide	Dose Kg.ai. h^{-1}	Time of application
Simazine	2.25-3.35	Pre -emergence
Diuron	2.62-3.40	Inter row spray
Paraquat	1% spray or 0.25-.0.4	Directed and spot sprays
Glyphosate	0.6-1.0	Directed and spot sprays
Glufosinate ammonium	0.3-0.5	Directed and spot sprays
Fusilade	25% ML	Directed and spot sprays

Rubber Plantation

It was reported that in 1876, a British explorer, Henry Wickham smuggled a large amount of seeds from Santarem area in Brazil to the Royal Botanical Garden (Kew Gardens) in London, England (Serier, 1993). Rubber seeds were distributed from there to the British colonies like Sri Lanka, Singapore, Malaysia, India and parts of Africa.

Weeds compete with young rubber seedlings for moisture and nutrients strongly enough to reduce latex yield. Tall weeds infesting new rubber plantations, can also shade young transplants. Weed competition and the competitiveness of weeds decline with tree age as the tree canopy intercepts more light and limits weed growth in the understory. Weeds hamper farm operations, such as pruning, fertilizer application, spraying, tapping and disease control (De Jorge, 1962, Wycherly, 1964). Weeds indirectly limit production by serving as host for organisms that are detrimental to rubber trees. Allelopathy may also adversely affect growth of rubber plants. Mature rubber trees are not always free from competition with weeds. Weedy vines such as *Mikania micrantha* and *Merremia vitifolia* if not controlled, will climb and cover the canopy of rubber trees.

Weed Flora

Merremia vitifolia, Ageratum conyzoides, Borreria hispida, Euphorbia hirta, Commelina benghalensis Cynodon dactylon, Paspalam conjugatum, Amaranthus spinosus, Cypreus kyllinga, Mikania micrantha, Chromolaena odorata Eleusine indica, Imperata cylindrica, Mitracarpus verticilitus, Ipomea triloba, Cyperus rotundus, Setaria palmifolia, Emilia sonchifolia, Artemisia vulgaris Conyza ambigua, Syndnedrella nodiflora, Crassocephalum crepidiodes, Mimosa pudica, Calapagonium mucunoides, Heliotropium indicum, Imperata cylindica, Sorghum halepense, Spermacoce alata, Rottboellia cochinchinensis, Digitaria ciliaris, Eragrostis sp, Frimbristylis sp, Murdannia nudiflora.

Mechanical/ Physical Weed Control

Tillage operations carried out between rows and planting intercrops help to reduce weed infestation, prevent soil erosion, provide soil cover and also improve soil quality. Mowing is also done to remove weeds from rubber plantations. However in tropical areas, weed growth is very fast making it necessary to repeat the process at biweekly intervals during rainy season. In many places weed biomass is gathered between rows and torched.

Cultural Methods

Cover cropping and intercropping is adopted in rubber plantations to modify the environment and make it less favourable for emergence and growth of weeds. Weed smothering legumes such as *Mucuna bracteate Mucuna pruriens, Pueraria phaseoloides, Centrosema pubescens , Calopogonium mucunoides* and *Calopogonium caeruleum* are some of the cover crops recommended for rubber plantations (Kobayashi et al. 2003).

Nursery

When seedlings are above two months old pre-emergence herbicides may be sprayed directed to the base of the plants. Herbicides like diuron (3kg ai.ha^{-1}) can control weeds upto 3 months (Mathew et al .1977). Directed spraying of herbicides can be done only if the rubber seedlings are arranged with gaps to allow passage of sprayer. Post emergence weed control can be done with paraquat in 3 months old plantations when the bark on the lower portion of the stem hardens. Care should be taken to direct the spray to the base of the seedlings without hitting the green tissues. Inter row spraying can be done using a single, drift reducing, flat fan spray nozzle. Paraquat can be substituted with glufosinate ammonium @ 0.4 to 0.8 kg.ha^{-1}

Immature Rubber Pantations

Weeds between rows can be controlled by slashing, mowing, herbicide application, intercropping with annuals or perennial crops or by planting leguminous cover crops. A mixture of Glyphosate (1.0kg ai. ha-1) + metsulfuron methyl (0.03kg ai .ha-1) can be sprayed at the base of rubber trees in less than 1 year old rubber plantations. Fluroxypyr, dicamba or picloram+2, 4-D can also be mixed with glyphosate and applied along with planting strip to keep it weed free (Faiz,2006).

Mature Rubber Plantations

Weeds in mature rubber plantations are removed either by regular slashing of small shrubs and weeds or by the application of herbicides such as glufosinate ammonium or paraquat at 3 to 6 months interval.

Table 3: Herbicides Recommended for Rubber Plantations

Herbicide	Dose Kg.ai. h^{-1}	Time of application
Dalapon	1-1.5%	Post- emergence
Simazine	2.25-3.35	Pre-emergence
Diuron	2.62-3.40	Inter-row sprays
Paraquat	0.1% spray or 0.25-0.4	Directed and spot sprays
Glyphosate	0.6-1.0	Directed and spot sprays
Glufosinate ammonium	0.3-0.5	Directed and spot sprays
Oxyfluorfen	0.12-0.25	Both pre and post emergence

Vegetables

A wide spectrum of grass and broad leaved weeds infest vegetables. This is especially true in tropics and subtropics where there is heavy rainfall and plenty of sunshine. To reduce use of chemicals in vegetable gardens an integrated approach will be more suitable.

Vegetables are normally grown in seed beds or furrows. A good tilth is necessary for healthy root growth and aeration of vegetable crop. Seed basins need regular irrigation. These factors also encourage weed growth in vegetable gardens.

Weed flora

Cyperus spp. *Cynodon dactylon, Panicum repens, Digitaria marginata, Chloris barbata, Brachiaria* spp. *Eragrostis* spp., *Dactyloctenium aegptium, Commelina benghalensis, Mimosa pudica, Achyranthus aspera, Phyllanthus niruri, Leucas apera, Celosia argentea, Polycarpaea corymbosa, Euphorbia hirta, Parthenium hysterophorus, Mollugo cerviana, Oldenlandia corymbosa.*

Methods of Weed Management

Vegetables are seasonal crops with a duration of 3 to 6 months. Seedlings are delicate and need at least two weeks of weed free period.

Physical and Mechanical Methods

In vegetables shallow cultivation is recommended with use of equipment's such as sweeps, knives, harrow, finger weeders or rotary hoes. This will help to loosen the soil for root aeration and also remove weeds from the root basins. Small weeds can be easily controlled by this method. It is effective in hot dry weather and in dry soils.

Cultural Method

In large scale cultivation of vegetables, mulching is widely used as it controls weeds and also improves productivity. Common mulches used are plastic film or organic matter such as straw, hay, coir pith etc.

Stale seed beds

Stale seed bed technique is adopted to reduce the density of weeds in the soil seed bank. Seed beds are prepared 2-3 weeks before planting. The beds are then irrigated to encourage maximum germination of weed seeds near the soil surface.

These weeds are removed before planting by either hand pulling or by use of non-residual post emergence herbicide to kill the germinated weeds just before or after planting. Vegetable seeds are then planted with minimum soil disturbance.

Treating the field with non-residual herbicides like glyphosate, paraquat or glufosinate ammonium just before planting or after planting before crop emergence is also effective.

Crop rotation is recommended to remove associated weeds from the site of cultivation. A good choice should include competitive crops such as legumes in the rotation.

Soil Solarisation

This is an eco-friendly weed control technique capable of removing a broad spectrum of weed species. It also affects soil properties and usually produces higher yield (Campiglia et al. 2000) in this technique transparent sheet (0.1 to 0.2mm) is used to cover the soil. The soil should be moist and the transparent sheet should be tucked into the soil in such a way that it is very well sealed. The warmer and summer months for 30 to 45 days. To get good results, the soil temperature must reach above 40°C to exert good effect on the weed seeds in the soil seed bank. After removing the plastic sheet, deep tillage should be avoided. It is more suitable for small areas of vegetable cultivation and also in green house conditions. The soil solarisation process has been mechanized for large scale cultivation of tomatoes.

Chemical Method

Table 4: Herbicides Recommended for Vegetable Crops

Herbicide	Dose g.ai.ha^{-1}	Time of application	Crop
Clomazone	0.18-0.	Pre-emergence	Pepper,cucumber
DCPA (Dacthal)	6.0-7.5	Pre-emergence	Onion, lettuce, cole crops
Metribuzin	0.15-0.5	Pre-emergence	Tomato
Napropamide	1.0-2.0	Pre-emergence	Tomato, pepper egg plant
Propachlor	5.2-6.5	Pre-emergence	Onion,cole crops
Trifluralin	1.0-1.5	Pre plant	okra
Glyphosate			All vegetables
Pendimethalin	0.59-1.44	Pre- or post emergence	All vegetables
Ethalfluralin	0.8-1.7	Pre-plant	Tomato, pepper, beans, squash

Sethoxydim	0.25-0.5	Post -emergence	
Loxinil	0.36	Post -emergence	Onion, garlic, leek
2,4-Dimethyl amine salt 58% SL	2.0	Post -emergence	Potato
Halosulfuron Methyl 75% WG	67.5	Post- emergence	Bottle gourd
Fernoxaprop-p-ethyl9.3%w/wEC	78.75	Post- emergence	Onion
Propaquizafop 10% EC	62.5	Post -emergence	Onion, Sugarbeet Cole crops
Quizalofop ethyl 5% EC	37.5-50.0	Post- emergence	Onion
Imazethapyr 35%+ Imazamox 35%	0.70	Pre-emergence, early post emergence	Cluster bean
Acetic acid products	5% -30%	Post emergence Spot application	Home gardens
Natural oil products	55% d-limonene or 70% citrus oil	Post emergence Spot application	Home gardens
Herbicide soaps	Ammoniated Soap of Fatty Acids. 22%	Post emergence Spot application	Home gardens
Corn gluten meal	9 kg per 92 sq m area	Pre-emergence	Home gardens

Orchards

India is the second largest producer of fruits and nuts. Both tropical and subtropical fruit trees grow well in our climate. While pineapple, mangoes and grapes are more popular in the south, strawberries, litchis, apples and citrus fruits are grown in North India and in the high ranges. Cultivation practices adopted for these fruit trees are different. However weed is a major concern in orchards. Weeds compete with trees of interest for growth factors like moisture and nutrients. Some weeds act as alternate host of diseases, nematodes or insect pests. They contribute to labour inefficiencies for harvest and discomfort from allergies.

Common orchard weeds

A wide range of weeds, broad leaved, grasses and perennial species grow in orchards. Their population is highly variable depending on several factors like soil type and the type of trees grown in the orchard.

Major weed flora

Bidens pilosa. Amaranthus viridis, Sonchus oleraceae, Tagetes minuta, Emex australis, Datura stramonium, Galinsoga parviflora, Solanum nigrum,

Oxalis spp., *Portulaca oleracea, Lolium mutiflorum, Digitaria sanguinalis, Echinochloa crusgalli, Cyperus rotundus, Sorghum halepense, Cynodon dactylon, Panicum repens, Eleusine indica, Pennisetum pedicellatum, Helicanthus elastic, Dendrophthoe falcata*

Mehods of Weed Management

Proper weed control improves quality and quantity of produce. It increases proper utilization of resources such as water and nutrients. It allows easy movement in the orchard and improves labour efficiency.

Cultural Method

Cultural practices adopted in orchards varies with the crop. In orchards where fruit trees such as mango, sapota and mangostene are grown, mowing and slashing reduces weed population.

In herbaceous fruit crops like strawberries and pineapple, mulching is a common practice. It reduces weed flora between rows and also improves productivity and fruit quality.

Cover cropping and growing intercrops is a common practice in mango and citrus orchards, Seasonal vegetables such as tinda, bottle gourd, bitter gourd, onion, chilies and cowpea are grown as intercrop, especially during initial stages of orchard establishment.

Torching is done in a controlled manner to reduce weeds and also to reduce pest and diseases.

Mechanical Control

Mechanical tools like jembes, spades, hoes, rakes, pangas, ploughs can be used in orchards for weed control in the interspaces. Use of mechanical methods may cause root injury to shallow root feeders. Wounds cause penetration of pathogens which lead to infection.

Electric weed control is an upcoming technology suitable for high value plants.

Chemical Method

In orchards, chemical weed control is the most popular method mainly due to ease of application, absence of mechanical damage and cost effectiveness.

Table 5: Herbicides Recommended for Orchards

Herbicide	Dose Kg.ai. h^{-1}	Time of application	Fruit crop
Dalapon	1 %	Post emergence	Mango, grapes, peaches
Simazine	2.0	Pre- emergence	Grapes
2,4-D Sodium salt	1-2	Post emergence	Citrus and grapes
Diuron	2.0	Pre -emergence	Pineapple, grapes, citrus, banana. Mango
Paraquat	3.0	Post -emergence	Mango, grapes
Glyphosate	3.0	Post -emergence	grapes
Oxyfluorfen	1.0	Post -emergence	Grapes, pineapple
Bromacil	3-4	Pre -emergence	Pineapple, citrus, apple

References

Ascard, J., Hatcher, P.E., Melander, B., Upadhyaya, M.K., 2007. Thermal weed control. In Upadhyaya, M.K., Blackshaw, R.E., 2007. Non-chemical weed management. Principles, Concepts, and Technology. CABI, London, UK.

Brainard, D.C. and Bellinder, R.R. 2004. Weed suppression in a broccoli–winter rye intercropping system. Weed Science 52, 281–290

Bond, W., Grundy, A.C., 2001. Non-chemical weed management in organic farming systems. Weed research 41, 383-405

Campiglia, E., Temperini, O., Roberto, M. and Saccardo, F. 2000. Effects of soil solarization on the weed control of vegetable crops and on the cauliflower and fennel production in the open field. Acta Horticulturae. 533(533):249-255

Chand, M., Singh, S., Bir, D., Singh, N. and Kumar, V. 2014. Halosulfuron methyl: a new post emergence herbicide in india for effective control of Cyperus rotundus in sugarcane and its residual effects on the succeeding crops. Sugar Tech 16 (1): 67-74

Das, T. K 2009.Weed Science-Principles and Applications. Jain Publishers

De Jorge, P. 1962. Report of botanical division. A Report by the Rubber Res. Inst. Malaya, 1961. pp.71–73.

Faiz, M.A.A. 2006. Efficacy of glyphosate and its mixtures against weeds under young rubber forest plantation. J. Rubber Res. 9: 50–60

Gupta, O. P. 2007. Weed Management- Principles and Practices. Agribios

Hamma, I. L. and Ibrahim, U. 2013 Weed Management Techniques of Horticultural Crops in Nigeria. American Eurasian J. Agric & Environ. Sci. 13(3):362-366

Harper, 1973. Efficiency of paraquat as weed control chemical for rubber nurseries. PANS 21: 401–405.

Kobayashi, H., Miura, S. and Oyanagi, A. 2004. Effects of winter barley as a cover crop on the weed vegetation in a no-tillage soybean. Weed Biology and Management 4, 195–205 (PDF) Cover crops and weed management. Available from: https://www.researchgate.net/publication/43265879_Cover_crops_and_weed_management

Mathew, M., Punnoose, K.I. and Potty, S.N. 1977. Report on the results of chemical weed controlexperiments in the rubber plantations in South India. J. Rubber Res. Inst. Sri Lanka 54: 478–488 (6) (PDF) Weed Management in Natural Rubber. Available from: https://www.researchgate.net/publication/342864007_Weed_Management_in_Natural_Rubber

Serier, J.B. 1993. History of Rubber. Desjonqueres Editions, Paris, 273 pp

Whycherly, P.R. 1964. Report of botanical division. A Report by the Rubber Inst. Malaya 1963. pp.51–21.

17

Modern Approaches in Weed Science and Management

Survival strategies of some plants help them to thrive in any adverse environmental conditions. We classify them as weeds or ruderals. Such weedy species have developed unique and effective physiological traits to withstand various stresses, and unique mechanisms to regulate growth under adverse environments. Climate change is a harsh reality observed in different parts of the world which brings new challenges in crop production and weed management.

Introduction of biotechnological tools in weed science and management has given new dimensions to weed research. Major biotechnological tools that have successfully contributed to weed management are herbicide tolerant crops, improvement of biocontrol agents and nanoherbicides, development of transgenic allelopathy in crops and characterization of weeds using molecular systematics. Currently biotechnological approach for weed management is gaining popularity in different parts of the world.

Molecular biology and genomics open new opportunities to explore weed physiology, biology, and genetics. Information gained from these fields will further our understanding of weed interaction. Genomics-based approaches will allow such traits to be characterized and manipulated so that some may be transferred into crop species, in such a way that a weed itself could be developed into a crop.

Introduction of Herbicide Tolerant GM Crops

Molecular biology has already staked its claim in weed management with the introduction of "Roundup Ready" crops which is the trademark for a line of genetically modified crop seeds resistant to glyphosate based herbicides patented by Monsanto. Roundup Ready soybeans was commercialized in 1996 followed by Roundup Ready corn in 1998.

These genetically modified (GM) crops showed resistance to glyphosate, a broad spectrum herbicide which could kill all other weeds in a cropped area. This increased the usage of glyphosate, but reduced the use of atrazine, metribuzin and alachlor which are highly soluble and can easily contaminate water sources.

This type of plant transformation has been possible by the transfer of cloned genes in susceptible plants through engineered vector technique. Currently among the herbicide resistant crops, glyphosate-resistant crops (Roundup Ready) represent more than 80% of the 120 million ha of transgenic crops grown annually worldwide (Duke et al., 2012). Main advantage of herbicide resistant crops is increased crop yield due to broader spectrum of weed control and reduced crop injury. Before the advancement of transgenic technology, traditional plant breeding tools were adopted to develop the first triazine resistant canola in 1984. The source of resistant trait was the weed *Brassica campestris* that had evolved resistance to triazine herbicide in the field. Recurrent backcrossing between the closely related weed and the crop led to stable introgression of the chloroplastic gene which was transmitted to the cytoplasm of the crop. The Indian Agricultural Research Institute (IARI) in 2021 has developed the first non GM herbicide tolerant rice variety Pusa Basmati (1979) and Pusa Basmati (1985). These varieties contain a mutated acetolactate synthase (ALS) gene making it possible for farmers to spray Imazethapyr, a broad-spectrum herbicide, to control weeds.

Table 1: Commercialized Herbicide Tolerant Crops

Crop	Herbicide	Transgenic	Year commercialized
Canola	Triazines	No	1984
	ALS inhibitors	No	1997
	Glufosinate	Yes	1995
	Glyphosate	Yes	1996
	Bromoxynil	Yes	2000
Com	ALS inhibitors	No	1992
	Sethoxydim	No	1996
	Glufosinate	Yes	1996
	Glyphosate	Yes	1998
Soybean	ALS inhibitors	No	1994
	Glyphosate	Yes	1996
Cotton	Bromoxynil	Yes	1995
	Glyphosate	Yes	1996
	Glufosinate	Yes	2005
Wheat	ALS inhibitors	No	2002
Rice	ALS inhibitors	No	2002
Sunflower	ALS inhibitors	No	2003
Alfalfa	Glyphosate	Yes	2005
Sugar beet	Glyphosate	Yes	2007

A1.S, acetolactate synthase.

Source: Data are from Green (2007, 2009).

Improvement of Biocontrol Agents and use of Nanotechnology

Bio herbicides used for the control of weeds contain microorganisms or products derived from microorganisms which are natural metabolites produced by these organisms in the course of their growth and development. Major constraints in the development of bio herbicides is to sustain the activity and efficiency of the organism in the formulation. Nano formulations are an emerging research area that has the potential for innovative developments such as controlled release of active phytotoxins derived from microbial sources as nano particles. The commercial products developed by combining these two technologies that combines the benefits of bioherbicides with nanotechnology in formulations can be expected to have acceptance in the market. However, further validation of the technology is required to ascertain that these nanoparticles can in no way harm the environment, since use of heavy metals in the development of nano particles is a source of concern especially in agricultural production.

Development of Transgenic Allelopathy in Crop Plants

With the emphasis on developing clean cultivation practices, attention has been focused to find suitable alternative strategies to reduce the load of chemical weed control in several crops. Identification of allelopathic interactions of crop plants by natural allelochemicals is considered as a potential means of reducing synthetic herbicides. Allelopathy is considered as an effective, economical and environment friendly weed management approach. It has been observed that allelopathic compounds released from leaves, flowers, seeds, stems and roots of living and decomposing plant materials can influence weed density and growth. Walnut and mango trees produce allelochemicals that suppress weed growth in their surroundings. Allelopathic cover crops such as rye and sorghum are widely used by organic farmers (Weston 1996, Khanh et al., 2005).

Traditional breeding techniques have not been successful in enhancing the allelopathic potential of crop plants, because production of secondary metabolite in the plant system could not be improved by such approaches. Production of allelochemicals can be enhanced by developing transgenic plants with foreign genes encoding for a particular weed-suppressing allelochemical, which would help to improve the competitiveness of the crop and suppress weeds in its vicinity. Enhanced production of allelochemical, sorgoleone in *Sorghum* was possible by using tools of molecular biology and transgenic approaches (Duke et al., 2001, 2003)

Cover crops can be used to suppress weeds, but the problem is that later on they tend to compete with the crop. Developing cover crops that will self-destruct after they have suppressed weeds, before they begin competing with the crop has been possible with transgene technology. A proof of this concept was demonstrated by Stanislaus and Cheng (2002) in *Nicotiana tabacum* (tobacco) by expressing the *Barnase* gene which encodes a highly toxic ribonuclease, which is under the control of a heat-shock promoter. With the onset of high temperature, the plants showed self-destruction. Environmental cues such as high temperature or day length or moisture content can serves as a signal for self-destruction.

Striga is a devastating parasitic weed that remains in the soil and causes damage throughout the world. In 2006, a novel strategy for control of *Striga* was introduced with the collaboration of BASF Corporation, CIMMYT (International Maize and Wheat Improvement Center), and the African Agricultural Technology Foundation.

The technology developed by them was called "Strigaway". Here, they used non-transgenic *Zea mays* (corn) which was resistant to ALS-inhibiting herbicides. The corn seeds were seed coated with the ALS-inhibiting herbicide, imazapyr. The germinating corn seedlings absorb some of the herbicide and subsequently transmits it to any attaching parasites, which are susceptible to the herbicide. The effectiveness of this technology has been demonstrated both in research and in local farmer trials (Groote et al., 2007). Major advantage of this technology is that the herbicide is already in the corn seed, so the farmer need not apply the herbicide.

Genomics for Trait Assessment

Molecular biology can also be used in other areas of weed science such as to gain a better understanding of how weeds compete and interact with neighboring plants, survive harsh environmental conditions, and evolve resistance to herbicides. The next generation of molecular biology tools, such as genomic resources, may yield novel weed management strategies and shed new light on what makes plants weedy.

Evolution of herbicide resistance can be studied by PCR (polymerase chain reaction)-based amplification and sequencing from genomic DNA which can give an insight into the mutation of the target sight that has contributed to herbicide resistance in plants (Tranel and Wright, 2002). Molecular markers can be used to understand dissemination of weed resistance. Such studies when coupled with studies of molecular basis for resistance, can be helpful in predicting and designing strategies to mitigate spread of herbicide resistance in plants. (Andrews et al., 1998, Rowe et al., 1997)

Application of genomic techniques in weed science has helped to decipher unique traits of weeds which imparts invasiveness, and adaptability to diverse environments. Modern high-throughput molecular and analytical tools offer exciting opportunities to study the development of herbicide resistance in weeds. Application of newer "*omics*" techniques such as proteomics, metabolomics, and physionomics in agriculturally important crops and weeds has helped to study crop-weed interactions. Since weedy traits are complex and there can be overlapping biological interactions, it is necessary to apply more than one "*omics*" platform to get a better understanding of the biological interactions (Maroli et al., 2018).

The use of omics techniques to study various biological aspects would provide greater opportunities to dissect the molecular and physiological mechanisms in developing resilient phenotypes. Utilising diverse *omics* platforms, functional genomics has seen rapid progress, resulting in a growing number of sequenced plant genomes and their corresponding expression repertoires. This has facilitated the development of plants selected for specific agronomic traits and biological processes. It has been opined that '*weedomics*' is the need of the hour to facilitate strategies for sustainable weed management practices and for innovations in crop improvement towards weed competitiveness (Tripathi, 2017).

Utilization of DNA microarrays for unravelling crop-weed interactions has been the beginning of molecular interventions for efforts towards developing weed resilient crop species (Lee et al., 2008). Of late, high-throughput genomics has aided the identification and characterization of key targets involved in herbicide response across different weedy species, which is critically important in the context of development of herbicide resistance in weeds (Shah et al., 2022). Innovative weed management strategies demand a convergence between conventional approaches like chemical weed control and biotechnological interventions to address the issues underlying herbicide resistance (Wong et al., 2022). There are recent efforts in understanding the role of short life cycle in weed ruderality in Brassicaceae, by genome assembly and QTL mapping in *Cardamine occulta* (Li et al., 2023). Initiatives such as the Earth BioGenome project, a joint effort of the International Weed Genomics Consortium, high throughput phenotyping and concerted field trials will be instrumental in addressing and mitigating the weed syndrome in crop science (Lewin et al., 2018; Exposito-Alonso et al., 2020; Li et al., 2022)

Use of this information can help to develop crop varieties or cultural strategies that allow crops to better compete with weeds, and to make better predictions of environmental changes and also develop more efficient and sustainable agriculture in future.

References

Andrews, T.S.,Morrison,I.N. Penner, G.A. 1998. Monitoring the spread of ACCase inhibitor resistance among wild oat (*Avena fatua*) patches using AFLP analysis. Weed Science.46:196-199

Basu, C, Halfhill, MD, Mueller, TC, Stewart, CN (2004) Weed genomics: new tools to understand weed biology. Trend Plant Sci 9:391–398 CrossRefGoogle ScholarPubMed

Belz, RG, Duke, SO (2014) Herbicides and plant hormesis. Pest Manag Sci 70:698–707 CrossRefGoogle ScholarPubMed

Berner DK, Ikie FO and Green, JM (1997) ALS-inhibiting herbicide seed treatments control Striga hermonthica in ALS-modified maize (*Zea mays*). Weed Tech 11: 704–707.

Boyette CD, Quimby PC, Connick WJ, Daigle DJ and Fulgham FE (1991) Progress in the production, formulation, and application of mycoherbicides. In: TeBeest DO (ed.) Microbial Control of Weeds. (pp. 209–224) Chapman and Hall, New York.

Chao, WS, Horvath, DP, Anderson, JV, Foley, ME (2005) Potential model weeds to study genomics, ecology, and physiology in the 21st century. Weed Sci 53:929–937 CrossRefGoogle Scholar

Dayan, FE, Duke, SO (2014) Natural compounds as next-generation herbicides. Plant Physiol 166:1090–1105 CrossRefGoogle ScholarPubMed

Duke, S.O., Scheffler, B.E., Dayan, F.E., Weston, L.A., Ota, E. 2001. Strategies for using transgenes to produce allelopathic crops Weed Technology.15:826-834

Duke, S.O. 2003. Weeding with transgenes Trends in Biotechnology 21: 192-195

Exposito-Alonso, M., Drost, H. G., Burbano, H. A. & Weigel, D. The Earth BioGenome project: opportunities and challenges for plant genomics and conservation. Plant J. 102, 222–229 (2020)

Gressel, J. 2000. Molecular biology of weed control. Transgenic Res 9, 355–382 https://doi.org/10.1023/A:1008946628406

Gressel J, Amsellem Z, Michaeli D and Sharon A (1997b) Synergized mycoherbicides for resistance management. In: De Prado R (ed.) Weed and Crop Resistance to Herbicides (pp. 267–274) Kluwer, Amsterdam

Groote, H., de Wangare, L., Kanampiu, F.2007. Evaluating the use of herbicide-coated imidazolinone-resistant (IR) maize seeds to control Striga in farmers' fields in Kenya. Crop Protection.26:1496-1506

Green, J.M.2007. Review of glyphosate and ALS-inhibiting herbicide crop resistance and resistant weed management. Weed Technology. 21:547-558

Green, J.M.2009. Evolution of glyphosate-resistant crop technology. Weed Science.57:108-117

Khanh, T.D., Chung, M.I., Xuan, T.D., Tawata,S.2005. The exploitation of crop allelopathy in sustainable agriculture production. Journal of Agronomy and Crop Science.191:172-184

Lee, R. M., & Tranel, P. J. (2008). Utilization of DNA Microarrays in Weed Science Research. Weed Science, 56(2), 283–289. http://www.jstor.org/stable/25148516

Lewin,H.A., Richar,S., Aiden.E.l and Zhang,G. 2022. The Earth BioGenome project 2020; Starting the clock.PNAS. Vol119(4) e2115635118 https://doi.org/10.1073/pnas.2115635118

Li, L.Z., Xu, Z.G., Chang, T.G., Wang, L., Kang, H., Zhai,D., Zhang,L.Y., Zhang,P., Liu,H., Zhu,X.G. and Wang,J.W.2022. Common evolutionary trajectory of short life-cycle in Brassicaceae ruderal weeds. https://doi.org/10.1038/s41467-023-35966-7

Maxwell BD, Roush ML and Radosevich S (1990) Predicting the evolution and dynamics of herbicide resistance in weed populations. Weed Tech 4: 2–13.

Maroli, A., Gaines, T., Foley, M., Duke, S., Doğramacı, M., Anderson, J., Tharayil, N. (2018). Omics in Weed Science: A Perspective from Genomics, Transcriptomics, and Metabolomics Approaches. Weed Science, 66(6), 681-695. doi:10.1017/wsc.2018.33.

Pfeifer TA and Griglilatti TA (1996) Future perspectives on insect pest management: Engineering the pest. J Invert Pathol 67: 109–119.

Rowe, M.L. Lee, D.J. Nissen, S.J., Bowditch, B.M., Masters, R.A. 1997. Genetic variation in North American leafy spurge (Euphorbia esula) determined by DNA markers. Weed Science.45:446-454

Robson PRH, McCormac AC, Irvine AS and Smith H (1996) Genetic engineering of harvest index in tobacco through overexpression of a phytochrome gene. Nat Biotechnol 14: 995–998.

Stanislaus Marisha A, and Cheng Chi-Lien. 2002. "Genetically Engineered Self-destruction: an Alternative to Herbicides for Cover Crop Systems." Weed science, v. 50 ,.6 pp. 794-801. doi: 10.1614/0043-1745(2002)050[0794:GESDAA]2.0.CO;2

Suh, H.S., Sato, Y.I. and Morishima, H. 1997. Genetic characterization of weedy rice (Oryza sativa L) based on morpho-physiology, isozymes and RAPD markers.

Saari LL, Cotterman JC and Thill DC (1994) Resistance to acetolactate synthase inhibiting herbicides In: Powles SB and Holtum JAM (eds) Herbicide Resistance in Plants. (pp. 83–139) Lewis, Boca-Raton.

Shah S, Lonhienne T, Murray C-E, Chen Y, Dougan KE, Low YS, Williams CM, Schenk G, Walter GH, Guddat LW and Chan CX (2022) Genome-Guided Analysis of Seven Weed Species Reveals Conserved Sequence and Structural Features of Key Gene Targets for Herbicide Development. Front. Plant Sci. 13:909073. doi: 10.3389/fpls.2022.909073

Tranel, P.J. and Wright, T.R. 2002. Resistance of Weeds to ALS-Inhibiting Herbicides: What Have We Learned? Weed Science, 50, 700-712. https://doi.org/10.1614/0043-1745(2002)050[0700:RROWTA]2.0.CO;2

Tripathi Niraj.2017. Weedonics a need of the hour.Current Science A Forthnightly Journal of Rsearch Vol112(2).416-417

Weston, V. C. M. 1999. The commercial realization of biological herbicides. Proc Brighton Crop Protection Conf, p. 281.

Wetzel, D.K., Horak, M. J. and Skinner, D.Z. 1999. Use of PCR-based molecular markers to identify weedy Amaranthus species. Weed Sci 47: 518–523.

Weston, L.A.1996. Utilization of allelopathy for weed management. Agronomy Journal.88:860-866

Wong ACS, Massel K, Lam Y, Hintzsche J, Chauhan BS. Biotechnological Road Map for Innovative Weed Management. Front Plant Sci. 2022 Apr 25;13:887723. doi: 10.3389/fpls.2022.887723. PMID: 35548307; PMCID: PMC9082642.

Annexure

Starts from Next Page

Approved uses of Registered Herbicides

Herbicide name & approved Crops	*Weed species*	Dosage /ha		Dilution In Water (Litres)	Waiting period / PHI between last application & harvest (days)
		a.i.(gm/ Kg)	Formulation in (gm/ ml / Kg/ ltr)		
Alachlor 50% EC (The use of Alachlor shall be completely banned with effect from the 31st December 2020)					
Cotton	*Digera arvensis* *Echinochloa colonum Eragrostis major* *Euphorbia hirta Phyllanthus niruri* *Portulaca oleracea* *Trianthema portulacastrum Flaveria australasica Gynandropsis pentaphylla*	2-2.5 kg	4-5 ltrs.	250-500	210-240
Maize	*Echinochloa colonum,* *Euphorbia hirta* *Eleusine indica*	2.5 kg	5 ltrs.	250-500	90
Groundnut	*Amaranths viridis* *Digitaria* spp. *Echinochloa* spp. *Euphorbia hirta Phyllanthus niruri Portulaca oleracea Trianthema portulacastrum Acanthosermum hispidum Flaveria australasica*	2.5 kg 1.5-2.5 kg	5 ltrs. 3-5 ltrs.	250-500 250-500	120-150 120-150
Soybean	*Amaranthus viridis* *Cleome viscose Cyperus iria Dactyloctenium aegyptium Echinochloa* spp. *Eleusine indiaca* *Setaria glauca*	2.5 kg	5 ltrs.	250-500	

Alachlor 10% GR (The use of Alachlor shall be completely banned with effect from the 31st December 2020)					
Cotton	*Dactyloctenium aegyptium*	2.0-2.5 Kg	20-25 Kg	-	-
Maize / Groundnut / Soybean	*Digitaria* spp., *Echinochloa* spp., *Chenopodium album*	1.5-2.5 Kg	15-25 Kg	-	-
Ametryne 80% WDG					
Sugarcane	*Dactyloctenium aegyptium*, *Digitaria sanguinalis*, *Cynodon dactylon*, *Ageratum conyzoides*, *Trianthema monogyna*, *Parthenium hysterophorus*	2.0 kg	2.5 kg	500	311
Anilofos 30% EC					
Transplanted paddy	*Echinochloa crusgalli* *Echinochloa colonum* *Cyperus difformis*, *Cyperus iria*, *Eclipta alba* *Ischaemum rugosum* *Fimbristylis* sp. *Marsilea quadrifoliata*	0.3-0.45 kg	1-1.5 ltrs.	375-500	30
Transplanted Paddy	*Echinochloa crusgalli* *Echinochloa colonum* *Cyperus difformis*, *Cyperus iria*, *Eclipta alba* *Ischaemum rugosum* *Fimbristylis* sp.	0.30-0.45 kg	1.66-2.5 kg	500-600	-
Anilophos 2 % G					
Transplanted rice	*Echinochloa crusgalli* *Echinochloa colonum* *Ischaemum rugosum* *Cyperus iria*, *Cyperus difformis*, *Fimbristylis* sp.	0.4-0.5 Kg	20-25 Kg	-	30

Atrazine 50% WP					
Maize	*Trianthama monogyna* *Digera arvensis, Echinochloa* spp *Eleusine* Spp. *Xantheium strumarium* *Brachiaria* sp, *Digitaria* sp, *Amaranthus viridis, Cleome viscose* *Polygonum* spp.	0.5-1.0 kg	1-2 kg	500-700	-
Sugarcane	*Protulaca oleracea,* *Digitaria* spp., *Boerhaavia diffusa,* *Euphorbia* spp., *Tribulus terrestris*	0.50-2.00 kg	1.00- 4.00 kg	500-700	
Azimsulfuron 50% DF					
Rice (Transplanted)	*Enchinochloa colonum,* *E. crusgalli, Cyperus* spp., *Fimbristylis miliacea, Ludwigia parviflora, Eclipta alba,* *Bergia capensis, Marsilea quadrifoliata, Ammania baccifera,* *Sphenoclea zeylanica*	35	70	300	59
Rice (Direct Seeded)	*Enchinochloa colonum,* *E. crusgalli, Cyperus* spp., *Fimbristylis miliacea, Ludwigia parviflora, Eclipta alba,* *Bergia capensis, Marsilea quadrifoliata, Ammania baccifera,* *Sphenoclea zeylanica*	35	70	300	59

Bensulfuron Methyl 60% DF					
Transplanted Rice. Pre-em (3 DAT)	*Marsilea quadrifoliata* *Eclipta alba, Ammania baccifera* *Ludwigia parviflora Sphenoclea* *Zeylenica , Monochoria vaginalis ,* *Alternanthera sessillis Cyperus iria ,* *Cyperus difformis , Fimbristylis* *miliacea, Scirpus royeli*	60 gm	100 gm	300 ltrs	88 days
Transplated Rice (post-em 20 DAT)	*Ammania baccifera* *Cyperus differmis Cyperus iria* *Eclipta alba* *Fimbristylis miliacca Ludwigia* *parviflora Marsilea quadrifoliata* *Monochoria vaginalis Alternanthera* *sessillis Scirpus royeli Sphenoclea* *zeylenica*	60gm	100 gm	300 ltrs.	71
Bentazone 480 g/l SL					
Soybean (Early POE: 2-3 leaf stage of weeds)	*Cyperus rotundus* *Achalipha indica Commelina* *bengalansis Echinocloa colanum* *Echinocloa crusgalli*	960	2000	500	62
Transplanted rice (Early POE: 2-3 leaf stage of weeds)	*Cyperus rotundus* *Cyperus diformis Ludwigia sps.* *Eclipla alba Echinocloa colanum* *Echinocloa crusgali*	960	2000	500	71

Bispyribac Sodium 10% SC					
Rice (Nursary) (10-12 DAS)	*Echinochloa crusgalli* *Echinochloa colonum*	20-25 gm	200 ml.	300 ltrs.	-
Rice(Transplanted) (10-14 DAP)	*Ischaemum rugosum* *Cyperus difformis, Cyperus iria*	20-25 gm	200-250 ml	300 ltrs.	78
Rice (Direct seeded) (10-15 DAS)	*Fimbristylis miliacea , Eclipta alba , Ludwigia parviflora, Monochoria vaginalis, Alternantheraphiloxeroides Sphenoclcea zeylenica*	20-25 gm	200-250 ml	300 ltrs.	78
Butachlor 50% EC					
Paddy (transplanted)	*Cyperus difformis* *Cyperus iria Echinochloa crusgalli, Echinochloa colonum, Eleusine indica, Eclipta alba, Fimbristylis miliacea, Ludwigia parviflora, Sphenoclea zeylanica*	1.25-2.00kg	2.5-4 ltrs	250-500	90-120
Butachlor 5% GR					
Transplanted Rice	*Echinochloa Crusagalli* *Digitaria sanguinalis* *Setaria* spp., *Commelina benghalensis, Fimbristylis milliacea, Cyperus iria,* *Eleusine indica, Panicum* spp., *Echinochloa Colonum, Eclipta alba, Cyperus Defformis, ludwigia paviflora.*	1.25 -1.87 Kg	25.00 –37.50 Kg	-	90 - 105

Butachlor 50 % EW					
Transplanted Rice	*Echinochloa colonum* *Echinochloa crusgalli, Cyperus difformis Cyperus iria* *Eclipta alba, Fimbristylis miliacea* *Ludwigia parviflora, Sphenoclea zeylanica Monochoria vaginalis*	1.25-1.5 Kg	2.5-3.0	2.50-500	-
Carfentrazone ethyl 40% DF					
Wheat (25-35 DAS)	*Chenopodium album,* *Melilotus Indica, Melilotus alba,* *Medicago denticulata, Lathyrus aphaca, Analgalis arvensis,* *Vicia sativa,* *Circium arvense, Rumex sp,* *Malwa* sp.	20gm	50 gm.	400	80
Direct seeded Rice (10-15 DAS)	*Ludwigia parviflora* *Digera arevensis Phyllanthus niruri Spilanthes sp, Eclipta alba Cyperus* sp.	25	62.50	300	102
Chlorimuron Ethyl 25% WP + Surfactant					
Soybean (3-15DAS)	*Cyperus rotundus* *Commelina benghalensis Celosia argentea Digera arvensis Cucumis trigonus Cyprus iria,* *Parthenium hysterophorus,* *Acalypha indica, Phyllanthus niruri,* *Trianthema portulacashurm,* *Caesulia auxillaris*	9 gm	36 gm.	300 ltrs. + non ionic surfactant 0.2 % (Iso-octyl phenoxyl-poloxetha nol 12.5%)	45

Rice (transplanted) (5-10 DAT)	*Echinochloa crusgalli, Eclipta alba, Commelina benghalensis, Chenopodium album, Cyperus rotundus, Echinochloa colonum*	6gm	24 gm.	500-600	60
Cinmethylin 10% EC					
Transplanted Rice	*Cyperus iria Fimbristylis milacea Monochoria vaginalis Commelina Benghalensis Echinocloa crusgalli Marsilea minuta*	75-100 gm	0.75-1.0 ltrs.	500-700	110
Clodinafop- propargyl 15%WP					
Wheat	*Phalaris minor* (Canary grass)	60gm	400 gm.	375-400	110
Clodinafop- propargyl 15% w/w DF					
Wheat	*Phalaris minor* (Canary grass)	60 gm	400	500	70
Clomazone 50%EC					
Soybean	*Digiteria sp. Echinochloa sp. Parthenium hysterophorus Commelina sp.*	0.75-1.00Kg	1.5-2.0 Ltrs.	500-600	90
Transplanted Rice	*Echinochloa crusgalli Echinochloa colonum Cyperus difformis Cyperus iria, Ludwigia parviflora, Eclipta alba*	0.4 - 0.5kg	0.8-1.0 ltr	500-750	90
Sugarcane	*Enchinochloa colonum Brachiaria repens Dactylotenium aegyptium Trianthema portulacastrum*	0.75-1.00 kg a.i./ha	1.5-2.0 ltr/ha	500 Lit	296

Cyhalofop Butyl 10% EC					
Rice (Directed seeded)	(*Echinochloa* spp.) Barnyard grass	75-80 gm	0.75-0.80ltr	500-600	90
2,4-D Dimethyl Amine salt 58% SL					
Maize	*Trianthema monogyna, Amaranthus sp., Tribulus terristeris, Boerhaavia diffusa, Euphorbia hirta, Portulaca oleracea, Cyperus sp.*	0.5 kg	0.86	400-500	50-60
Wheat	*Chenopodium album, Fumaria parviflora, Melillotus alba, Vicia sative, Asphodelus tenuifolius, Convolvulus arvensi,*	0.5-0.75 kg	0.86-1.29	500-600	-
Sorghum	*Cyperus iria, Digera arvensis, Convolvulus arvensis, Trianthema sp., Tridax procumbens, Euphorbia hirta, Phyllanthus niruri.*	1.8 kg	3.1	500-600	-
Potato	*Chenopodium album, Asphodelus tenuifolius, Anagalis arvensis, Convolvulus arvensis, Cyperus iria, Portulaca oleracea.*	2.0 kg	3.44	400	-
Sugarcane	*Cyperus iria Digitaria sp. Dactylactenium aegyptium Digera arvensis Portulaca oleracea Commelina benghalensis Convolvulus arvensis*	3.5	6.3	500	-

Aquatic Weeds Non crop area	*Eichhornia crassipes.* *Parthenium hysterophorus, Cyperus rotundus*	0.5-1.0 kg 2.65 kg 2.5 kg	0.86-1.72 4.56 4.30	600-700 300-400 300-400	15-20 15-20 -
2,4-D Sodium salt Technical (having 2,4-D acid 80 % w/w) (Earlier Registered as 80%WP)					
Citrus	*Euphorbia* spp. *Convolvulus arvensis* *Coronopus didymus* *Amaranthus viridis* *Oxalis corniculata* *Tribulus terrestris* *Fumaria parviflora* *Sonchus arvensis*	1.00-2.5 kg	1.25-3.2 kg	600	>6 months
Grapes	*Convolvulus* spp. *Tridax procumbens*	2.0	2.5	500	> 90 days
Maize	*Amaranthus viridis,* *Trianthema portulacastrum* *Phyllanthus niruri.,* *Euphobia geniculata,* *Amaranthus spinosus.* *Cleome chelidonii,* *Lagasca mollis*	1.00 Kg.	1.25	500	120(Pre-em) 90(post-em)
Sugarcane	*Boerhaavia diffusa* *Chenopodium album* *Tribulus terristris* *Portulaca oleracea* *Xanthium* spp. *Convolvulus arvensis Amaranthus spinosus Digera arvensis Celosia argentina.*	2.0-2.6	2.5-3.25	600-900	300

Wheat	*Leucas aspera, Chenopodium album, Vicia sativa, Argemone maxicana, Fimbristylis miliacea, Anagalis arvensis, Amaranthus spinosus.*	0.5-0.84 kg.	0.625-1.0	500	90
Aquatic Weeds	*Boerhaavia hispada, Eichhornia crassipes.*	1.5 kg	1.85.	600-1000	-
Non crop land	*Parthenium hysterophorus,* *Cyperus rotundus,* *Solanum elaeagnifolium.*	2.5-6.0 kg. 4-8 Kg 1.8 kg	3.2-7.5 5-10 2.25	600-1000 500-600 500-600	- - -
2,4-D Ethyl Ester 38 % EC (having 2,4-D acid 34 % w/w)					
Maize	*Trianthema monogyna*, *Amaranthus* sp., *Portulaca oleracea., Tribulus terristris, Boerhaavia diffusa, Euphorbia hirta, Cyperus* sp.	0.9 kg	2.65 ltr	400-450	50-60
Sorghum	*Cyperus iria, Striga* sp. *Digera arvensis, Convolvulus arvensis, Trianthema* sp., *Tridax procumbens, Euphorbia hirta, Phyllanthus niruri.*	1.0 kg	2.94	425	-
Transplanted Paddy	*Echinochloa colonum, Echinochloa crusgalli.*	0.85 kg	2.5	400	-

Wheat	*Chenopodium album, Asphodelus tenuifolius,*	0.45-0.75 kg	1.32-2.2	450-500	-
	Fumaria parviflora Melilotus alba. Spergula arvensis				
Sugarcane	*Cyperus iria, Digitaria* sp., *Dactyloctenium, Aegyptiana, Digera arevensis, Portuluca oleeracea, Commelina benghalensis, amaranthus* sp., *Convolvulus arvensis*	1.2 to 1.8	3.53- 5.29	500	300-330
Aquatic Weeds	*Eichhornia crassipes*	2.5 kg	7.5	700-1000	-
2,4-D Ethyl Ester 4.5 % GR (having 2,4-D acid 4 % w/w)					
Transplanted Rice	*Echinochloa Coloum E. Crusgalli Panium ischaemum Cynodon dactylon (germinating) Cyperus rotundus (germinating) Cyperus iria C. difformis Ludwigia parviflora Monochoria Vaginalis Marsilea quadrifoliata Cyanotis cucutata Eclipta alba Ammania baccifera*	1.0 kg	25 kg	-	-

Diclofop Methyl 28% EC					
Wheat	*Avena fatua,* *Phalaris minor*	0.7-1.0 kg	2.5-3.5 ltr	500	90
Diuron 80% WP					
Cotton	*Amaranthus* spp, *Chenopodium album,* *Convolvulas arvensis* *Setaria glauca,* *Digitaria* sp, | *Portulaca oleracea,* *Xanthium strumerium,* *Anagallis arvensis,* *Asphodelus temifolius,* *Euphorbia* sp, *Visia sativa* *Paspalum conjugatum,*	0.75-1.5 kg	1-2.2Kg.	625	-
Banana	*Cyperus iria,* *Commelina benghalensis,* *Digitaria* sp, *Amaranthus* spp, *Dactyloctenium, Chloris barbata, Eragrostis zeylenica,*	1.60 kg	2 kg.	625	-
Rubber	*Grasses & Non grasses*	1.6-3.2 kg	2-4kg.	625	-
Maize	*Cyperus iria,* *Echinochloa* spp, *Digitaria* spp, *Chenopodium album,* *Eleusine sp, Amaranthus* sp, *Phyllanthus niruri*	0.8 kg	1.0 kg.	600	-

Citrus (sweet orange)	*Cyperus iria*, *Tribulus Terristris*, *Digera arvensis*, *Commelina nudiflora*, *Cocumis trigonus*.	2-4.0kg	2.5-5.0kg	600	-
Sugarcane	*Cyperus iria*, *Portulaca racea*, *Echinochloa rusgalli*, *Cynotis* spp, *Amaranthus* spp, *Convonvulus* Spp., *Digitaria* spp.	1.6-3.2kg	2.0-4.0 kg.	600	-
Grapes	*Cleome viscose*, *Chenopodium album*. *Cyperus iria*, *Euphorbia hirta*, *Alternanthera echinata*, *Amaranthus* spp, *Argemone maxicana*, *Ipomoea* spp, *Xanthium strumerium*, *Fumeria parviflora*,	1.6kg	2.0 kg.	625	-
	Asphodelus tenuifolius, *Medicago denticulata*, *Eleusine aegyptia*.				
Diclosulam 84% WDG					
Soybean	*Cyperus* spp, *Commilena benghalensis*, *Euphorbia geniculata*, *Digera arvensis*, *Acylipha* spp, *Echinochlo colona*	22-26gm	26.2-30.9 Time of application 0-3 DAS	500	60

Groundnut	*Amaranlhus viridis*, *Parthenium hysterophorus*, *Trianthema* sp., *Euphorbia geniculate*, *Cyperus* spp., *Echinochloa colona*	22-26	26.2-30.9	-	77
Ethoxysulfuron 15% WDG					
Transplanted Rice.	*Fimbristylis miliacea* *Cyperus iria* , *Cyperus difformis*, *Scirpus* sp., *Eclypta alba*, *Marsilea quadrifoliata*, *Ammania baccifera*, *Monochoria vaginallis*	12.5-15gm	83.3-100gm	500	110
Fenoxaprop-p-ethyl 9.3% w/w EC (9% w/v)					
Soybean	*Echinochloa colonum*, *Echinochloa crusgalli*, *Digitaria sp*, *Eleusine indica*, *Setaria sp*, *Brachiaria* sp.	100gm.	1111 ml. (15-20 DAS)	250-300	100
Rice (transplaned)	*Echinochloa crusgalli*, *Echinochloa colona*	56.25 gm	625 ml. (10-15 DAT)	300-375	70
Blackgram	*Echinochloa crusgalli*, *Echinochloa colona* *Digitaria* sp. *Dactylocteneum* *Aegyptium*	56.25-67.5 g	625-750ml. (15-20 DAS)	375-500	43

Cotton	*Echinochloa* sp. *Eluesine indica* *Dactylocteneum Aegyptium Eragrostit minor*	67.5 g	750ml. (20 -25 DAS)	375-500	87
Onion	*Echinochloa colonum* *Dactyloctenium aegyptium*	78.75	875	375	10
Groundnut	*Echinochloa* sp.	78.75	875	375	89
Fenoxaprop-p-ethyl 10% EC					
Wheat	*Phalaris minor*	100-120gm	1.0-1.20 kg	250-300	110
Fenoxaprop-p-ethyl 6.7% w/w EC					
Rice (Transplanted & Direct Seeded)	*Echinochloa* sp.	56.6-60.38g	812.5-875	375-500	61
Fluazifop-p-butyl 13.4% EC					
Soybean	*Echinochloa colonum, Echinolchloa crusgalli, Eleusine indica, Cyanodon dactylon, Dactyloctenium Aegyptium, Digitaria* sp., *Setaria* sp.	125-250 g	1000-2000	500	90

Flucetosulfuron 10% WG					
Rice (Transplanted)	*Echinochloa colonum* *Echinolchloa crusgalli* *Digitaria sanguinalis* *Paspalum discichum* *Paspalum scrobitulatum* *Leersia hexandra* *Panicum repens* *Setaria glauca* *Dinebra retroflexa* *Cyprus difformis* *Cyprus iria* *Fimbristylis miliaceae* *Alternanthera philoxeroides*	25	250	500	90
	Alternanthera sessilis *Marsilea quadrifolia Ammania baccifera Eclipta alba* *Eclipta prostrate Monochoria vaginalis Lindernia ciliate Ludwigia parviflora Sphenoclea zeylanica Commelina diffusa Cyanotis axillaris*				

Fluchloralin 45% EC					
Cotton	*Acanthospermum hispidum,* *Cleome viscosa,* *Datura* sp. *Trianthema monogyna* *Tridax procumbens, Cynodon dactylon (germinating)* *Amaranthus* spp., *Portulaca* spp, *Achyranthus aspera, Euphorbia hirta, Cenchrus cathorticus,* *Digitaria sanguinalis,* *Eleusine sp,* *Panicum sp,* *Lagasca mollis,* *Gynandropsis pentaphylla,* *Achalypha indica*	0.9-1.2kg	2.0-2.68 ltrs.	500-800	180
Soybean	*Eragrostis* sp., *Boerhaavia hispada, Cyperus compestris,*	1.0-1.5kg.	2.22-3.33	500-800	120-150
Flufenacet 60% DF					
Paddy (Transplanted)	*Echinochloa crusgalli* *Echinochloa colonum* *Cyperus iria*	120 gm	200 gm	500	90-110
Flumioxazin 50% SC					
Soybean	*Commelina benghalensis,* *Digera arvensis, Euphorbia* spp., *Phyllannthus niruri, Echinochloa crusgalli*	125 g.a.i/ha	250ml/ha	500	110

Wheat	*Runnex* spp., *Medicago denticulate, Coronopus didymus, Chenopodium album, Phalaris minor,* *Avena fatua*	125 g.a.i/ha	250 ml/ha	500	137
Fluthiacet Methyl 0.3% CE					
Soybean	*Commelina sp,* *Digeru arvensis, Acalypha indica, Amaranlhus viridis.*	13.6	125	500	73
Glufosinate Ammonium 13.5% SL (15% w/v)					
Tea	*Panicum repens,* *Borreria hispida, Imperata cylindrical, Digitaria sanguinalis, Commelina benghalensis, Ageratum conyzoides, Eleusine indica, Paspalum conjugatum*	375-500 gm	2.5-3.3	375-500	15
Cotton	*Echinochloa* sp. *Cynodon dactylon Cyperus rotundus Digitaria marginata Dactylocteneum aegyptium*	375-450 gm	2.5-3.0	500	96
Glyphosate 20.2% SL IPA salt					
Non Crop area	*Phyllanthus niruri,* *Ageratum conyzoides, Parthenium hysterophorus, Sorghum halepense, Amaranthus spinosus, Alternanthera sessilis, Cynodon dactylon,* *Cyperus rotundus, Echinochloa colonum, Trianthema portalucastrum*	0.82-1.23 kg	4.1-6.15	400-500	N/A

Glyphosate Ammonium salt 20 % SL					
Non Crop area	*Cynodon dactylon* *Commelina benghalensis Panicum* spp. *Dactyloctenium aegyptium* *Eragrostis major* *Poa anua* *Cyperus rotundus Parthenium hysterophorous Acalypha indica* *Digeria arvensis Phyllanthus niruri* *Euphorbia geniculate* *Corchorus actangularis Saccharum spontenium Eleusine indica Imperata cylindrical Ageratum conzoides*	4.52-6.79g a.i./litre	20-30ml/lit	300-600	-
Glyphosate 41% SL IPA Salt					
Tea	*Arundinella bengalensis* *Axonopus compressus Cynodon dactylon Imperata cylindrical Kalm grass* *Paspalum scrobiculatum* *Polygonum perfoliatum*	0.820-1.230kg.	2.0-3.0	450	21
Non-cropped area	*Soghum helepense and other dicot & monocot weeds in general*	0.820-1.230kg.	2.0-3.0	500	-
Glyphosate 54% SL (IPA Salt)					
Non Crop Area	*Ageratum conyzoides* *Alternenthera sessilis Commilina spp* *Cyperus spp Echinochloa* sp. *Eclipta alba Iscaemum rogosum Setaria spp*	1.8 kg	3.33 ltrs.	400-500	-
Glyphosate Ammonium Salt 5% SL					

Tea	*Ageratum conyzoides Biden pilosa Boreria latifolia Cynodon dactylon Cyperus rotundus Digitaria sanguinalis Euphorbia* spp. *Imperata cylendrica Paspalum conjugatum*	1.5 kg.	30 ltrs.	500	7 days
Non Crop area	*Cynodon dactylon Cyprus rotundus Digera arvensis Digitaria sanguinalis Eragrostis minor Euphorbia* spp. *Parthenium hysterophorus Tribulus terrestris Xantrhium stremerium*	2 kg.	40 ltrs.	500	-
Glyphosate 71% SG (Ammonium Salt)					
Tea & Non Crop area	*Acalypha indica Ageratum conyzoides Cychorium intybus Digera arvensis Cynondon dactylon Cyperus rotunedus Digitaria sanguinalis Eragrostis* spp. *Ipomea digitarea Paspalum conjugatum Sida aculata*	2.13 kg	3.0 kg.	500	7
Halosulfuron Methyl 75% WG					
Sugarcane	*Cyperus rotundus*	60-67.5	80-90	375	294
Maize	*Cyperus rotundus* *Cyperus iria*	67.5	90	375	45

Bottle gourd	*Cyperus rotundus* *Cyperus iria*	67.5	90	375	46
Haloxyfop R Methyl 10.5% w/w EC					
Soybean	*Brachiaria* sp. *Digitaria sanguinalis* *Dinebra arabica* *Echinochloa* sp.	108-135	1000-1250	500	60
	Eleusine indica *Eragrostis* sp. *Pnicum isochmi*				
Imazethapyr 10% SL					
Soybean	*Cyperus difformis* *Echinochloa colonum E. crusgalli Euphorbia hirta Croton sperrsifeorus, Digera arvensis, Commelina Benghalensis*	100 gm	1.0 Ltr.	500-600	75
Groundnut	*Cyperus difformis* *Commelina benghalensis, Trianthema portulacasturm, Eragrostis pilosa*	100-150 gm	1.0-1.5 ltrs.	500-700	90
Imazethapyr 10% SL + Surfactant					
Soybean (1-2 Leaf stage of weeds or 7-14 days after sowing)	*Echinochloa colonum* *Brachiaria mutica, Euphorbia hirta Commelina benghalensis Dinebra arabica, Digitaria* spp.,	75-100gm+ MSO adjuvant @ 2ml/l of water	750-1000 ml+ MSO adjuvant @ 2ml/l of water	375	72
Groundnut (1-2 Leaf stage of weeds or 7-14 days after sowing)	*Echinochloa colonum* *Euphorbia hirta Commelina benghalensis Digera arvensis, Amaranthus viridis, Physalis minima.*	100-150 gm+ MSO adjuvant @ 2ml/l of water	1000-1500 ml+ MSO adjuvant @ 2ml/l of water	375	102

Imazethapyr 70% WG + Surfactant					
Soybean (2-3 leaf stage of weeds)	*Cyperus routandus Echinochloa* spp. *Dinebra arabica Digera* spp., *Brachiaria mutica, Commelina benghalensis Commelina communis Euphorbia geneculata Cyanotis axiallaris*	70 g/ha + Surfactant (Cyspread) @ 1.5ml/Litre + Ammonium Sulphate @ 2 g/lit of Water	100 g/ha + Surfactant (Cyspread) @ 1.5ml/Litre+ Ammonium Sulphate @ 2 g/lit of Water	500	56
Isoproturon 50% WP					
Wheat	*Phalaris minor Avena fatua Poa annua*	1.0kg	2.0	750	-
Isoproturon 75% WP					
Wheat	*Phalaris minor Avena fatua Poa annua*	1.0kg	1.33 kg.	750	0 days
MCPA, Amine salt 40% WSC					
Transplanted Rice	*Cyperus rotundus Impmoea reptans Ammania baccifera Lippia nodiflora Alternanthera* sp. *Ludwigia parviflora Marsilea quadrifoliata*	0.8-2.0 kg	2-5	400-600	

Wheat	*Chenopodium album,* *Asphodelus tenuifolius Fumaria parviflora Carthamus oxyacantha Launea* sp., *Pluchia lanceolata, Melilotus indica, Vicia hirsuta, Lathyrus aphaca, Medicago denticulata, M. lupulina, Spergula arvensis, Argemone maxicana, Phyllathus niruri.*	1.0 kg	2.5	300-600	
Metamifop 10% EC					
Direct seeded Rice	*Barnyard grass (Echinocloa spp), Sacchiolepis Dactyloctenium, Digiteria, panicum*	100 g.a.i	1000 ml	350	87
Metamitron 70% SC					
Sugarbeet	*Sedges & Grasses Cynodon dactylon Cyperus rotundus Dactyloctenium aegyptium Broad Leaves Convolvulus arvensis Chenopodium album Parthenium hysterophorus Digera arvensis*	a) 2-3 leaf stage of weed – 0.7 kg a.i/ha, b) 4-6 leaf stage of weed – 1.4 kg a.i/ha, c) 8-10 leaf stage of weed – 1.4 kg a.i/ha	a)2-3 leaf stage of weed – 1kg/ha, b) 4-6 leaf stage of weed – 2 kg/ha, c) 8-10 leaf stage of weed – 2 kg/ha	500	90
Methabenzthiazuron 70% WP					
Wheat (PE –2DAS)	*Phalaris minor,* *Avena fatua,* *Avena ludoviciana, Poa annua,*	1.05-1.4kg	1.5-2.0 kg.	700-1000	100
Wheat (Post –EM 30 DAS)	*Polypogom monspliensis, Anagallis arvensis, Chenopodium album*	1.05-1.75kg	2.0-2.5 kg.	700-1000	100

Wheat (Early POE.16-18 DAS)	*Phalaris minor,* *Avena fatua,* *Avena ludoviciana, Chenopodium album*	0.7-0.87 kg	1.0-1.25 kg.	700-1000	100
Metolachlor 50% EC					
S loybean	*Echinochloa colonum* *Eleusine indica Digitaria* sp. *Dactyloctonium aegyptium Panicum* sp. *Cyperus* sp. *Amaranthus viridis*	1.0 kg	2.0 ltrs.	600-750	-
Metribuzin 70% WP					
Soybean	*Digitaria* spp. *Cyperus esculentus Cyperus campestiris Borreria* spp. *Eragrostis* spp.	0.35-0.525 kg	0.5-0.75kg.	750-1000	30
Wheat	*Phalaris minor* *Chenopodium album* *Melilotus* spp.	Medium soil-0.175kg Heavy soil - 0.21kg	0.25 kg 0.30 kg.	500-750	120
Metsulfuron Methyl 20% WP					
Wheat	*Chenopodium album,* *Melilotus indica, Lathyrus aphaca, Anagallis arvensis, Vicia sativa, Cirsium arvense.*	4 gm	20 gm	500-600 + Surfactant (Iso-Octyl Phenoxyl-Poloxethano l 12.5%)@ 500 ml/ha	80

Rice (transplanted)	*Cyperus rotundus, Spheanochlea* spp., *Fimbristylis* sp. *Ludwigia parviflora Marsilea quadrifoliata*	4 gm.	20 gm.	500-600	60
Sugarcane	*Cyperus esculentus, Amaranthus viridis, Portulaca oleracea, Parthenium hysterophorus, Trianthema* sp., *Cleome viscosa, Solanum* sp., *Commelina benghalensis, Euphorbia* sp., *Digeria* sp.	6	30	500-600 (Add non - ionic surfactant Iso-octyl-phenoxyl - poloxethanol 12.5% @ 2ml per liter of spray volume (0.2%)	346
Metsulfuron Methyl 20% WG					
Wheat	*Chenopodium album Melilotus indioca Melilotus alba Lathyrus aphaca Anagalis arvensis Vicia sativa Rumex denticulate Convolvulus arvensis Meedicago denticulate*	4 gm.	20 gm.	500-600 + Surfactant (Iso-Octyl Phenoxyl-Poloxetha nol 12.5%) @0.2%	76
Transplanted Rice	*Monochoria vaginalis Ludwigia parviflora Ludwigia adscendens Marselea quadrifoliata Eclipta alba Oxalis minima Dapatorium juncecum Commelina benghalensis Ammania baccifera Sphenoclea zeylanica Caesulia axillaries.*	4 gm	20 gm.	500-600+ Surfactant (Iso-Octyl Phenoxyl-Poloxetha nol 12.5%) @0.2%	71
Orthosulfamuron 50% WG					

Transplanted Rice (Paddy)	*Echinocloa* spp. *(Barnyard grass)* *Cyperus* spp. *(Nut grass)* *Scirpus* spp. *Ludwigia parviflora (water crest)* *Fimbristylis* spp. *(Hoora grass)* *Rotala* spp.	60-75	150 3 DAT	500	65
Oxadiargyl 80% WP					
Transplanted Rice	*Echinochloa crusgalli* *E. Colonum, Cyperus iria, C. difformis,* *Eclipta alba,* *Ludwigia quadrifoliata*	100	125	500	97
Sunflower	*Echinochloa colonum* *Dactyloctenium aegyptium*	240	300	500	81
Oxadiargyl 6%EC					
Transplanted Rice	*Echinochloa crusgalli* *Echinochloa colonum,*	100gm	1.66 ltrs	500	97
Cumin	*Cyperus iria, cyperus difformis, Eclipta alba Ludwigia quadrifoliata Chenopodium album* *Remex* sp., *Melilotus indica, Asphodelus tenuifolius*	60-75gm	1.0-1.25 ltrs.	500	87
Mustard	*Chenopodium album,* *Melilotus sp*	90	1500	500	35

Oxadiazon 25% EC					
Transplanted Rice	*Echinochloa crusgalli E. colonum Cyperus iria C. difformis* *Marsilea quadrifoliata,* *Eclipta alba, Ludwigia* sp.	0.5kg	2.0 ltrs.	500	-

Oxyflourfen 0.35% GR					
Rice (Direct sown puddled or Transplanted)	*Echinochloa* sp. *Cyperus difformis Cyperus iria* *Eclipta alba* *Ludwigia parviflora Fimbristlylis miliacia, Marsilea spp*	100-150 gm	30-40 kg.	-	-
Oxyflourfen 23.5% EC					
Rice (Direct sown as pre-emergence)	*Echinochloa* sp. *Cyperus iria, Eclipta alba,*	150-240 gm	650-1000	500	-
Tea	*Digiteria,* *Imperata, Paspalum, Borreria hispida,*	150-250 gm	650-1000	500-750	15 days
Onion	*Chenopodium album,* *Amaranthus viridis,*	100-200 gm	425-850	500-750	-
Potato	*Chenopodium ,* *Coronpus Trianthema, Cyperus,* *Heliotropium*	100-200 gm	425-850	500-750	-
Groundnut	*Echinochloa colonum* *Digitaria arginata*	100-200 gm	425-850	500-750	-
Mentha	*Echinochloa colona,* *Cyperus* spp., *Solanum nigrum,* *Amaranthus viridis, Sphenochlea* sp., *Anagallis arvensis, Chenopodium album, Commelina benghalensis,* *Digitaria sanguinalis, Eclipta alba,* *Euphobia* spp., *Ludwigia parviflora,*	206	904.3	500	10
	Portulaca spp.				

Pendimethalin 30% EC					
Wheat	*Phalaris minor,* *Chenopodium album, Melilotus alba, Portulaca oleracea, Anagallis arvensis, Fumaria parviflora, Poa annua*	Light soil-1.0 kg, Medium soil-1.25 kg, Heavy soil-1.5 kg	3.3 ltr. 4.2 5.0	500-700 500-700 500-700	-
Rice (Transplanted &direct sown Upland)	*Echinochloa colona,* *E. crusgalli, Fimbristylis miliacea, Marselia quadrifoliata, Alternanthera sessilis, Ammonia baccifera, Ludwigra parviflora, Eclipta alba,* *Cyperus difformis*	Light to Heavy soil 1-1.5kg	3.3 –5 Ltrs.	500-700	
Cotton	*Echinochloa* spp. *Euphorbia hirta Amarnanthus viridisPortulaca oleraceaTrianthema* spp. *Eleusine indica*	0.75-1.25kg	2.5-4.165 ltrs	500-700	150
Soybean	*Echinochloa* spp., *Euphorbia* spp., *Amarnanthus viridis, Portulaca oleracea, Trianthema* spp., *Eleusine indica*	0.75-1.0kg	2.5-3.3 ltrs.	500-700	110
Pigeon pea	*Digitaria sanguinalis* *Digera arvensis Amaranthus* sp. *Euphorbia hirta Trianthema* sp. *Cyperus* sp. *Eragrostis* sp.	0.7 – 1.00	2.5 – 3.33	500	133

Pendimethalin 5 % G					
Rice (Transplanted & Direct sown puddled)	*Echinochloa colona, E. crusgalli, Fimbristylis miliacea, Marselia quadrifoliata, Alternanthera sessilis, Ammonia baccifera, Ludwigra parviflora, Eclipta alba, Cyperus difformis*	1.0-1.5 kg	20-30 kg	-	-
Pendimethalin 38.7% CS					
Soybean	*Echinochloa colonum Dinebra arabuica Digitaria sanguinalis Bracharia mutica Dactyloctinum aegyptium Portulaca oleracea Amaranthus viridis Euphorbia geniculata Cleome viscose*	580.5-677.25gm	1500-1750	500	40
Cotton	*Digitaria sanguinalis, Echinochloa colonum, Dinebra arabiaca, Brachiaria mutica, Eragratis minor, Portulaca oleracea, Commelina communis, Amaranthus* spp., *Parthenium hysterophorus*	580.5-677.25gm	1500-1750	500	101
Chilli	*Echinochloa colonum, Dinebra arabiaca, Brachiaria mutica, Amaranthus* spp., *Portulaca oleracea, Commelina* spp., *Parthenium hysterophorus, Digera arvensi, Physelis minima*	580.5-677.25gm	1500-1750	500	98

Onion	*Dactyloctinum aegyptium* *Digitaria sanguinalis, Echinochloa* sp., *Dinebra Arabic, Portulaca oleracea, Euphorbea geneculata, Commelina* sp., *Digera arvensi, Amaranthus viridis, Trianthenut portulacastrum*	580.50-677.25gm	1500-1750	500	104
Groundnut	*Echinochloa colonum, Digitaria marginata, Commelina bengalensis, Portulaca oleracea*	580.5-677.25gm	1500-1750	375	103
Mustard	*Chenopodium album,* *Digera arvensis, Amaranthus species*	338.625 gm	875	375-400	111
Cumin	*Portulaca oleracea,* *Digitaria* spp, *Digera arvensis*	580.5-677.25 gm	1500- 1750	375-500	91
Pinoxaden 5.1% EC					
Wheat	*Phalaris minor (Canary grass)* *Avena ludoviciana (Wild oat)*	40-45 g	800-900 ml 30-35 DAS	225-300	90
Penoxsulam 21.7 % SC					
Rice (Transplanted)	*Ammania bacifera,* *Cyperus difformis, Echinochloa colonum, Echinochloa crusgalli, Cyperus iria, Fimbristylis miliacea, Ludwigia* spp. *Monochoria* spp. *Sphenecelea zeylanica,*	22.5 to 25 (pre- emergence 0-5 DAT) 20 to 22.5 (post-emergence 10-12 DAT)	93.7 to 104.2 83.3 to 93.7		60

Penoxsulam 2.67% OD						
Rice (Transplanted Rice)	Grasses	*Echinochloa Colona* *Echinochloa Crusagalli*	22.5-25	900-1000	300-500	60
	Sedges	*Cyperus difformis*				
	Broad Leaved Weeds	Caesulia axillaris		ml/ha		
Pretilachlor 37%EW						
Transplanted Rice		*Echinochloa crusgalli* *Echinochloa colonum Cyperus difformis Cyperus iria* *Digitaria sanguinalis* *Fimbristylis miliacae Eclipta alba* *Ludwigia parviflora* *Monochoria vaginalis*	0.60-0.75 kg	1.5-1.875 ltrs.	500	90
Pretilachlor 30.7% EC						
(Direct seeded rice under puddled condition)		*Echinochloa crusgalli* *Echinochloa colonum Cyperus difformis Cyperus iria*	0.45-0.60kg.	1.5-2.0 ltr.	500	110
Pretilachlor 50% EC						
Transplanted Rice		*Echinochloa crusgalli* *Echiniochloa colonum Cyperus difformis Cyperus iria Fimbristylis miliacae Eclipta alba* *Ludwigia parviflora Monochoria vaginalis Leptochloa chinensis* *Panicum repens*	0.50-0.75 kg.	1.0-1.5 ltrs.	500-700	75-90
Propaquizafop 10% EC						
Soybean		*Echinochloa colonum, Echinochola crusgalli, Digiteria sanguinalis, Dactyloctenium eigyptium, Eleucine indica*	50-75 g	500-750	500-750	21

Blackgram	*Echinochloa colonum, Echinochola crusgalli, Digiteria sanguinalis, Dactyloctenium eigyptium, Eleucine indica*	75-100 g	750-1000	500-750	21
Onion	*Echinochloa colonum, Digiteria sanguinalis, Dactyloctenium eigyptium, Phalaris minor*	62.5	625	500	7
Paraquat dichloride 24% SL					
Tea (Post-emergence directed inter row application at 2-3 leaf stage of weeds)	*Imperata* *Setaria* sp., *Commelina benghalensis,* *Boerraria hispida,* *Paspalum conjugatum,*	0.2-1.0 kg	0.8-4.25 ltr (For season long weed control, use 2.5-5.0 ltr for initial application. For subsequent repeat spot application use 1 litre)	200-400	Not Necessary (For season-long weed control, muse 2.5 to 5 lit for initial application. For subsequent repeat spot application use 1 lite)
Potato (Post-emergence overall / inter-row application at 5-10 % emergence)	*Chenopodium* sp. *Angallis arvensis Trianthema monogyna Cyperus rotundus Fumeria parviflora*	0.5 kg	2.0 ltr.	500	100
Cotton (Post-emergence directed inter row application at 2-3 leaf stage of weeds)	*Digera arvensis,* *Cyperus iria, Trianthema monogyna,* *Corchorus* spp., *Leucas aspera, Euphorbia* spp.	0.3-0.5 kg	1.25-2.0	500	150-180

Rubber (Post-emergence directed inter row application at 2-3 leaf stage of weeds)	*Digitaria* sp., *Eragrostis* sp., *Fimbristylis* sp.	0.3-0.6 kg	1.5-2.5	600	N.A.
Coffee	*Digitaria marginata, paspalum Conjugatum, Ageratum, Conyzides, Borreria hispida, Euphorbia hirta, Commelina benghalensis, Eleusine indica*	250	1.0	400	N.A.
Rice [pre-plant (minimum tillage) before sowing/ transplanting for controlling standing weeds]	*Echinochloa crusgalli, Cyperus iria, Ageratum conyzides, Commelina benghalensis, Marsilea quadriofoliata, Brachiaria mutica*	0.3-0.8 kg	1.25-3.5	500	N.A.
Wheat [pre-plant (minimum tillage) before sowing]	*Grassy & Broad leaf weeds*	1.0 kg	4.25 ltrs	500	120-150
Maize [pre-plant (minimum tillage) before sowing]	*Cyperus rotundus, Commelina benghalensis, Trianthema monogyna, Amaranthus* sp., *Echinochloa sp*	0.2-0.5 kg	0.8-2.0 ltrs	500	90-120
Maize (Post-emergence directed inter row application at 2-3 leaf stage of weeds)	*Cyperus iria, Cyperus rotundus, Commelina benghalensis Amaranthus* sp. *Echinochloa sp Trianthema monogyna*	0.2-0.5 kg	0.8-2.0 ltrs	500	90-120
Grapes (Post-emergence directed inter row application at 2-3 leaf stage of weeds)	*Cyperus rotundus Cynodon dactylon Convolvulus* sp. *Portulaca* sp. *Tridax* sp.	0.5 kg.	2.0ltrs.	500	90

Apple (Post-emergence directed inter row application at 2-3 leaf stage of weeds)	*Rosa moschata* *Rosa eglantaria* *Rubus ellipticus*	0.75 kg	3.25 ltrs	700-1000	N.A.
Aquatic weed control Water ways Canals, Ponds etc	*Eichhornia crassipes* *Hydrilla* *Typha latifolia*	1000 1000 1000-2000	4.25 4.25 4.25-8.5	600-1000 600 600-1000	N.A
Pyroxasulfone 85% w/w WG					
Maize	*Echinochloa crusgalli, Eleusine indica, Phyllanthus niruri*	127.5	150	500	103
Wheat	*Phalaris minor*	127.5	150	500	131
Soybean	*Echinochloa colonum, Celosia argentia, Trianthema porulacastrum, Amarthanas viridis, Digeria arvensis*	127.5	150	500	94
Pyrazosulfuron Ethyl 10% WP					
Transplanted Rice	*Cyperus Iria, Cyperus difformis, Fimbristylis miliacea, Monochoria vaginalis, Ludwigia parviflora*	10-15 g	100-150	500-600	95

Pyrithiobac Sodium 10% EC					
Cotton (Gossypium)	*Trianthema Spp* *Amaranthus Spp Chenopodium Spp* *Digera Spp Celosia argentia*	62.5-75 gm	625-750	500	160
Pyrozosalfuron Ethyl 70% WDG					
Transplanted Rice	*Echinicloa* spp, *Cyparus rotundus, Ludwigia parviflora*	21g	-	-	43

Quizalofop-ethyl 5% EC					
Soybean	*Echinochloa crusgalli* *E. colomum* *Eragrostis* sp.	37.5-50 gm.	0.75-1.0	500-600	95
Cotton	*Echinolchloa crusgalli, Echinochloa colonum, Dinebra retroflexa Digiteria marginata*	50.5	1000	500	94
Groundnut	*Echinochloa colonum,* *Dinebra retroflexa* *Dactyloctenium* sp.	37.5-50.0	750-1000	500	89
Black gram	*Eleusine indica,* *Dactyloctenium aegyptium, Digitaria sanguinalis, Eragrostis* sp., *Paspalidium* sp., *Echinochloa* sp., *Dinebra ratroflexa*	37.5-50.0	750-1000	500	52
Onion	*Digitaria* sp., *Eleusine indicia, Dactyloctenium aegyptium, Eragrostis* sp.,	37.5-50.0	750-1000	375-450	7
Quizalofop-ethyl 10% EC					
Soyabean	*Love grass (Eragrostisi ipilosa),* *Crab grass (Digitaria sanguinalis/* wild finger/ Makra grass Viper grass,	375-45.0	375-450	300-500	69-103
	Barnyard grass, *sanwa/Samel,* Brown top millet				

Quizalofop –p-tefuryl 4.41% EC					
Soybean	*Echinochloa* spp. *Dinebra arabica Digitaria sanguinalis Cynodon dactylon Hemarthria compressa Eleusine indica*	30-40 gm	750-1000 ml	400	30
Sulfentrazone 39.6% w/w SC					
Soybean	*Acalypha* sp., *Commelina* sp., *Digera* sp., *Cyprus* sp. *Echinochloa* sp., *Brachiaria* sp., *Dinebra* sp.	360	750	500	88
Sugarcane	*Trianthema* sp., *Digera* spp., *Amaranlhus* spp., *Phyllanthus* spp., *Euphorbio* spp., *Dacteloctenium* spp., *Digitaria* spp., *Brachiaria* spp., *Echinocloa* spp., *Cynodon* spp, *Cyperus* spp.	720	1500	500	306
Sulfosulfuron 75% WG					
Wheat	*Phalaris minor* *Chenopodium* sp. *Melilotus alba*	25 gm	33.3 gm	200-250 + Cationic surfactant 1250ml/ha	110
Tembotrione 34.4% SC					
Maize	*Trianthema portulacastrum, Echinochloa* sp. *Bracheria* sp.	120g	286ml	500L	55

Triallate 50% EC					
Wheat	Avena fatua	1.25 kg	2.5 kg.	250-500	150

Triasulfuron 20% WG					
Wheat	*Chenopodium album, Anagallis arvensis, Medilotus alba, Rumex* spp, *Medicago denticulata, Fumeria pomiflora, Cronopus didymus, Spergula arvensis Malvela perviflora*	20	100	500	81
Tea	*Ageratum conyzoides, Borreria* spp., *Crassocephalum crepidiodes, Oxalis* spp., *Bidens pilosa, Conyza ambigua, Drymaria diandra, Emillia sonchifolia, Mitracarpus verticilatus, Syndnedrella nodiflora.*	25	125	500	7
Topramezone 336 g/l w/v SC					
Maize	*Elusine indica, Digitaria sanguinalis, Dactyloctenium aegyptium, Echinocloa* spp., *Chloris barbata, Parthenium hysterophorus, Digera arvensis, Amaranthus viridis, Physalis minima, Alternanthera sessilis, Convolvulus arvensis, Celotia argentea.*	25.2 to 33.6 g a.i./ha + MSO adjuvant @ 2 ml/l of water	75 to 100 ml + MSO adjuvant @ 2 ml/l of Water	375	83

Herbicide Combinations

Anilofos 24% +2,4-D ethyl Ester 32% EC					
Transplanted rice	*Echinochloa crusgalli Echinochloa colonum Ischaemum rugosum Fimbristylis miliacea*	(0.24+0.32) to (0.36 + 0.48) kg	1-1.5 ltrs.	300	90

Bensulfuron methyl 0.6%+Pretilachlor 6% GR					
Transplanted Rice	*Echinochloa crusgalli,* *Echinochloa colonum, Cynodon dactylon Cyperus iria, Cyperus difformis, Cyperus rotundus, Fimbristylis miliacea, Ludwigia parviflora, Marselia quadrifolia, Enhydra fluctuans, Sphenoclea zeylanica, Eclipta alba, Ammania baccifera.*	60 + 600 gm	10 kg	N.A.	88
Carfentrazone ethyl 0.43% + Glyphosate 30.82% EW					
Tea	*Ageratum conyzoides* *Bidenspilosa Borreria* sp. *Crassocephalumcr epidioides Cynadon* sp. *Cyperous* sp. *Digitaria* sp. *Eleusine indica Mimosa* sp. *Mltracarpus villosus* *Oxalis* sp.	12.90 +924.60	3000	500	7
Non-cropped area	*Ageratum conyzoides* *Axonopus* sp. *Brachiaria* sp. *Commelina* sp. *Cynodon dactylon Cyperous* sp. *Digitaria* sp. *Eleusine indica Imperata cylendrica Lantana camera Parthenium* sp.	12.90 +924.60	3000	500	--

Carfentrazone ethyl 20% + Sulfosulfuron 25% WG					
Wheat	*Phalaris minor* *Avena ludoviciana Chenopodium album Medilotus alba* *Rumex spp*	20+25 +750 ml Surfactant	100	300	110
Clodinafop Propargyl 15% + Metsulfuron Methyl 1% WP					
Wheat	*Phalaris minor,* *Avena fatua, Chenopodium album, Melilotus* sp., *Fumaria parviflora, Vicia sativa,* *Rumex* sp., *Anagallis arvensis, Coronopus didymus,* *Lathyrus* sp., *Convolvulus arvensis*	60+4	400	375 (Add 1250 ml surfactant at the time of sparying)	100

Clodinafop propargyl 9% + Metribuzin 20% WP (W/W)					
Wheat	*Phalaris minor* *Chenopodium album, Melilotus sp* *Vicia sativa, Rumex sp Medicago sp Cronopus didymus* *Dinebra vetroflexa*	54+120	600	300	120
Clomazone 20%+2,4-D EE 30% EC					
Transplanted Rice	*Echinochloa colonum,* *Echinochloa crusgalli, Cyperus iria,* *Cyperus difformis, Eclipta alba,Leptochloa chinensis,* *Panicum repens, Fimbristylis miliacea, Marsilea* *quadrifoliata, Ludwigia parviflora.*	0.250-0.375 Kg	1.25 ltrs.	500	100-110
Fenoxaprop-p-ethyl 7.77% w/w + Metribuzin 13.6% w/w EC					
Wheat	*Phalaris minor* *(Little seed canary grass) Chenopodium album (Lambs quarter) Lathyrus aphaca* *(Meadow Pea)* *Rumes* Sp. *(Golden dock) Melilotus* spp.*(Sweet clover)* *Avena ludoviciana.*	100+175	1250	375	110
Fluazifop-p-butyl 11.1% w/w + Fomesafen 11.1% w/w SL					
Soybean	*Echinochloa colona* *Digitaria sp Eleusine indica Dactyloctenium aegyptium Brachiaria reptans* *Commelina benghalensis Digera arvensis* *Trianthema* sp. *Phyllanthus niruri Aclypha indica* *Dinebra arbica*	250	1000	500	71

Groundnut	*Echinochloa colona* *Digitaria* sp. *Eleusine indica* *Dactyloctenium aegyptium* *Commelina benghalensis Eluropus villosus* *Indigofera glandulosa Chloris barbata Trianthema* sp. *Digera arvensis Cleome viscose* *Phyllanthus niruri Amaranthus virdis* *Cyperus* sp.	250	1000	500	82
Hexazinone 13.2% + Diuron 46.8 % WP					
Cane	*Enchinochloa colonum* *Dactylotenium aegyptium Trianthema monogyna* *Amaranthus virdis* *Ipomea spp Cyperus rotundus Cyperus esculentus* *Setaria spp Parthenium hysterophorus Euphorbia hirta*	1200 gm (264+936)	2 Kg	500	282-306
ndaziflam 1.65% w/w (2%w/v) + Glyphosate Isopropylammonium 44.63% w/w (40%w/v) SC					
Tea	*Ageratum* sp. *Borreria* sp. *Eleusine indica*	50+1000 to 70 + 1400 g.a.i/ha	2500 to 3500 ml/ha	500 L	14 days
Imazethapyr 35% + Imazamox 35% WG					
Soybean	*Echinochloa Colonum,* *Dinebra Arabica, Digitaria sanguinalis, Brachiaria mutica Commelina benghalensis Euphorbia hirta*	70 g a.i/ha + MSO Adjuvant @ 2 ml/l of water	100 g MSO Aadjuvant @ 2ml/l of water	375-500	56

Groundnut	*Echinochloa Colonum,* *Digira arvensis, Commelina benghalensis Euphorbia hirta Amaranthus viridis Physalis minima Brachiria* Spp. *Trianthema portulacastrum*	70 g a.i/ha + MSO Adjuvant @ 2 ml/l of water	100 g MSO Aadjuvant @ 2ml/l of water	375-500	83
Cluster bean	*Echinochloa colonum,* *Euphorbia* spp., *Digitaria arvensis, Amaranthus viridis.*	70 g a.i./ha + MSO adjuvant @ 2m/l of water	100 g/ha + MSO adjuvant @ 2m/l of water	500	64
Red gram	*Euphorbia* spp., *Amaranthus viridis.*	70 g a.i./ha + MSO adjuvant @ 2m/l of water	100 g/ha + MSO adjuvant @ 2m/l of water	375-500	125
Mesoulfuron Methyl 3% + Iodosulfuron Methyl Sodium 0.6% WG					
Wheat	*Phalaris minor* *Medicago denticulata Chenopodium album* *Melilotus* sp. *Rumex* sp. *Anagallis arvensis Coronopus didymus* *Lathyrus aphaca Fumaria parviflora*	(12+2.4 gm)	400 ml.	400-500 + Surfactant (Genopol LRO fluid) @ 500 ml/ha	96
Mesotrione 2.27% w/w + Atrazine 22.7% w/w SC					
Maize	*Trianthema* spp., *Cyperus* spp. *Digitaria sanguinalis, Echinochloa* spp. *Dactyloctenium aegyptium*	875 gm	3500	500	42
Sugarcane	*Trianthema* spp. *Amaranthus viridis, Echinochloa colona,* *Digitaria sanguinalis Cyperus rotundus,* *Dactyloctenium aegyptium*	875 gm	3500	500	190

Metsulfuron Methyl 10% + Chlorimuron ethyl 10% WP					
Transplanted Rice (Pre-emergence application-3 DAT	*Cyperus iria,* *Cyperus difformis, Fimbristlylis miliaceae, Eclipta alba,* *Ludwigia parviflora, Cyanotis axillaries,* *Monocoria vaginalis, Marsilea quadrifoliata,*	4gm	20 gm.	300	90
Metsulfuron Methyl 10% + Carfentrazone ethyl 40% DF					
Wheat	*Rumex dentatus* *Rumex spinosus Medicago denticulate Malva parviflora Lathyrus aphaca Chenopodium album* *Melilotus alba Melilotus indica Anagallis arvensis* *Solanum nigrum* *Vicia sativa* *Convolvulus arvensis*	25	50	300	100
Oxyflurofen 2.5% + Glyphosate (Isopropyl amime salt)41% SC(w/w)					
Tea	Ageratrum Conyzoids Cyperous sp Borreriabispida Pospalumcon jugatum Digitaria ciliaris	50+820	2000	500L/ha.	14
Pendimethalin 30%+ Imazethapyr 2% EC					
Soybean	*Echinocloa crusgalli* *Digera arvensis Commelina benghalensis,* *Amaranthus viridis Portulaca oleracea*	(750+50) to (900+60) gm	2.5-3.0 ltrs	500-600	90
Penoxsulam 0.97% w/w + Butachlor 38.8% w/w SE					
Transplanted Rice	*Echinochloa colonum,* *Echinochloa crusgalli, Cyperus iria,* *Cyperus difformis, Marsilia quadrifoliata,* *Alternanthera* spp.	820 g.a.i/ha	2000ml/ha	750	60

Penoxsulam 1.02 % + Cyhalofop-butyl 5.1% OD					
Rice (Direct seeded Rice)	*Echinochloa colona* *Echinochloa crusgalli Leptochloa chinesis Eleusine indica Alternanthera sessilis Caesulia axillaris Cyperus* spp	120-135	2000-2250	300-500	60
Rice (Transplanted Rice)	*Echinochloa* colona *Echinochloa crusgalli Leptochloa chinesis Caesulia axillaris Cyperus difformis Cyperus* spp	120-135	2000-2250	300-500	60
Pendimethalin 35% + Metribuzin 3.5% w/w SE					
Wheat	Canary grass (*Phalaris minor*), Wild Oat (*Avena ludoviciana*), Lamb's quarters (*Chenopodium album)*, Field bindweed (*Convolvulus arvensis)*, Swine Cress (*Coronopus didymus*), Burmuda grass (*Cyanadon doctylon*)	875 +87.5 to 1050 + 105 gm	2.5-3.0	500	123
Pendimethalin 38.4% + Pyrazosulfuron ethyl 0.85% ZC					
Transplanted Rice	*Echinochloa* colona (wild rice), *Echinochloa* crus galli (banyyard grass), Marsilea quadrifolia (comrnon water clover), *Ludwigia parvifl* ora (rvater crest), *Cyperus dffirmis* (comrnon sedge/, *Cyperus* spp.	900 +20	2000	375	86

Pretilachlor 6% + pyrazosulfuron Ethyl 0.15%(H)					
Paddy	Grassy weeds, Broad Leave, Sedges	600+15	10	-	83
Pretilachlor 6.0% +Pyrazosulfuron Ethyl 0.15% GR					
Transplanted Paddy	*Echinochloa Colonum* *Echinochloa Crusagalli Ludwigia paviflora,* *Elipta alba,* *Leptochola chinensis, Monochoria vaginalis,* *Cyperus difformis, Cyperus iria, Fimbristylis miliaceae*	600	10	-	83
Pretilachlor 30.0% + Pyrazosulfuron Ethyl 0.75% WG					

Triafamone 20% + Ethoxysulfuron l0% WG					
Rice Transplanted	*Echinochloa colona,* *Echinochloa crusgalli, Cyperus rotundus,* *Cyperus difformis, Fimbistylis miliaceae,* *Marsilea quadrifolia.*	44+22.5	225	300	83
Rice Direct Seeded	*Echinochloa* colona, *Cyperus rotundus*, Digera arvensis, Commelina benghalinsis.	44+22.5	225	300	83

Colour Plates

Cosmos

Calotropis gigantea (L.)R.Br.

Tridax procumbens L.

Wings, *Acer marophyllum* Pursh

Baloon *Physalis minima* L.

Feathery persistent styles: *Sphaeranthus indicus* L. (Indian globe thistle)

Dispersal by Wind: *Vernonia cinerea* (L.)Less

Dispersal by Water: *Monochoria vaginalis* (Burm.f.) Presal ex Kunth

Pappus hairs, *Emilia sonchifolia* (L.) DC.

Aquatic Weeds and Their Management

Removing aquatic weeds for paddy cultivation in Kole lands of Kerala

Thypha latifolia

Limnocharis flava

Pistia stratiotes (Source: Wikipedia)

Eichhornia crassipes

Salvinia molesta (Source: Wikipedia)

Marsilea quadrifolia (Source: Wikimedia)

Cabomba caroliniana

Hydrilla verticillata

Parasitic Weeds

Orobanche aegyptiaca *Striga asiatica*
(*Source*: CABI Compendium)

Cassytha filiformis
Source: CABI Compendium

Cuscuta reflexa (*Source*: Schwabe india)

Viscum capitellatum

Viscum angulatum

Dendrophthoe falcata

D. neelgherrensis

Helixanthera wallichiana

Scurrula parasitica

Index